BSI: the story of standards

by C Douglas Woodward

with a Foreword by
Lord Sherfield, GCB, GCMG
President of BSI

London
British Standards Institution
1972

Published by the British Standards Institution,
2 Park Street, London W1A, 2BS

Printed by Gaylard and Son Limited, London SE14

ISBN: 0 580 07591 5

Contents

Author's preface

For ten years, from 1955 to 1964, I had the good fortune to be entrusted with the task of making the idea of British Standards better known to industry and the public at large. It was totally absorbing job and quite easy in that this was a time when BSI was developing from being a 'back-room' organization to the national force it has become today.

It was greatly flattering when my two chiefs of those days, Roy Binney and Gordon Weston – themselves personifying, one could say, a large part of the story of British Standards – invited me to write this history, and I can only hope that they – and other readers – will find it as interesting to read as it has been to write.

Roy Binney, and the present Director-General, I must specially thank for helping to bring me up-to-date on standards business. There are others to whom I am particularly obligated: the author of an earlier history of BSI, whose material I have unashamedly cannibalized; Margaret Hardie, another old colleague from whom I received a wealth of notes; the greatly missed Rosalind Mockridge who steered the project through its early stages; Richard Townsend who has been tireless in getting my manuscript through its various committee stages!

I hope that this account of yet another inspired British invention – imitated all over the world – will help to dispel any lingering confusion as to what British Standards are and do.

C D W

Foreword

The British Standards Institution issued the first British Standards in 1901. Its activities and their importance to British industry have expanded steadily since it was founded.

In 1951 a history of BSI was published to mark the 50th anniversary of the Institution. Since that time, BSI has had to keep pace with the gathering momentum of technological advance and throughout the period this work has been carried on under the able direction of Mr Roy Binney, who retires from BSI at the end of this year. This is, therefore, an opportune moment to supplement the previous history with an account of the last twenty years.

In 1901 there was little public concern for the consumer or the environment. Electronics and other modern technologies were unknown. With technological progress the public becomes better informed and develops a demand for higher standards of performance, of safety, of reliability, and of quality, which only a strong national standards organization can provide.

Seventy years ago, Britain enjoyed a dominant position in world trade, and the acceptability of British Standards was unchallenged in other countries. Today, Britain is only one of many advanced industrial societies, but we are on the threshold of entry into the EEC, potentially the largest trading area in the world. Through BSI, Britain is playing a leading part in the development of internationally accepted standards and the removal of non-tariff barriers to trade. BSI works through the International Organization for Standardization and the International Electrotechnical Commission on a world scale, and through a network of committees which is specifically concerned with Europe. These bodies provide admirable channels for international co-operation and are unusually free from political pressures.

The rapidity of technological change demands constant review of objectives and procedures of work on standards, and lends urgency to it. Under the leadership of Mr Robert Feilden, since 1970 Director General of BSI, an appraisal of the priorities for the 1970s is being made. The task of shaping the policies for the future is one in which the members of BSI – that is, the whole of British industry – can and should take an active part through the committees set up for this purpose.

This book is designed to stimulate interest in standards work and to encourage financial and administrative support for it. I hope its publication will lead to a better understanding of the complexity and urgency of this work, and of its importance to our national prosperity and welfare.

LORD SHERFIELD GCB GCMG
President of BSI

Introduction

Making to measure: the logic of standards

Today, the economic logic of standardization is universally accepted as being fundamental to the achievement of an improved standard of living for all, and the existence of standards is often taken for granted. Yet nationally agreed standards, now so obvious, and the formation and growth in Britain of the first standards body in the world, did not come until the beginning of the present century. Why was the 'standards idea' so late in developing?

The answer lies in the unique circumstances of the later stages of the industrial revolution. The momentum of technological advance made long-term technical agreements between manufacturers hard to formulate or to keep, and the comparatively localized patterns of trade did not require reference to national standards. Above all, there was a prevailing spirit of competition, suspicion and jealousy. Fear of undercutting and imitation perhaps constituted the major barrier to the achievement of agreement on the adoption of national standards. Against this background, the formation in 1901 of the British Engineering Standards Association, forerunner of the British Standards Institution, represented not only a triumph of common sense but a major breakthrough in the history of technology and industrialization.

Once the discipline of standards has been accepted, there are enormous advantages to be reaped in terms of time saved in design and acceptable variety reduction in type and size ranges. Sir John Wolfe Barry, whose efforts were largely responsible for the formation of the BESA, speaking on the occasion of a dinner held in his honour in the Goldsmiths' Hall in 1911, pointed out that 'The loss of time, energy and money, for want of systematic standardization prior to 1902 would have been recognized as immense if its amount could have been known.'

In recognition of the enormous value of standards to the national economy, the Government has consistently supported this work. But it has not sought to control or unduly influence it, for such control would undermine the democratic basis of all standards work in this country. BSI operates on principles laid down in its Royal Charter of 1929 (supplemented in 1931 and 1968). These place on BSI the task of achieving a consensus between producers and users. (see page 12). Participation in, and support for this work is entirely voluntary, and the co-operation and goodwill of all parties is fundamental. Recognition of this principle makes the discipline of standards generally acceptable, without the compulsion that would be ultimately self-defeating. An exception to this general rule of voluntary acceptance arises in cases where reference is made to British Standards in legislation. When statutory authorities decide to make regulations, usually in the interests of health, safety or consumer protection, it is convenient for law enforcement and for the public to call up technical criteria in the form of a published standard.

Today, some sixty countries have their own national standards organizations. It is a measure of Britain's pioneering role that many of these have modelled their operations on those of the British Standards Institution.

The operation of BSI

The principle of the voluntary acceptance of standards influences all the Institution's operations and procedures. These must evolve as the demands on BSI develop, and must take into account that the majority of the work is dependent upon those whose time and service is freely given. Subscribing members total nearly 15 000 at present, and include companies; professional, research and trade and consumer organizations; educational bodies; local authorities; nationalized industries; consultant and professional engineers. The value of their subscriptions may vary from a few pounds to many thousands, according to the size of the organization and the use that it makes of British Standards.

Committee members, numbering some 25 000, are nominated to serve on BSI's councils and committees by professional or trade associations, Government authorities, local and national, or other interested bodies because of their specialist qualifications, knowledge and special interests including that of the general community. It is the intention that every committee should thus comprise a balance of viewpoints of manufacturers and users and the general community interest. From these are also chosen those who will present the UK viewpoint as delegates participating in international standards negotiations. The service of committee members is entirely voluntary and represents an 'invisible' contribution of several millions of pounds each year to the Institution's work.

The income of BSI derives from members' subscriptions, a Government grant, sales of publications, and fees for certification and other services. The overall budget currently amounts to around £3 million per annum. This is divided under three separate heads: standards making (£2.5M), quality assurance (£0.35M) and Technical Help to Exporters (£0.26M). Proposals for new standards work may come from any responsible body, including established BSI committees. BSI looks for clear evidence of support for a project from all the major interests involved, and assesses the value of each project in the light of the national need, in terms of economic factors, such as exports, or other considerations such as safety. The necessary resources for carrying it out must also be available.

Responsibility for all major policy decisions is vested in the Executive

BRITISH STANDARD SECTIONS

ISSUED BY

The Engineering Standards Committee.

SUPPORTED BY

THE INSTITUTION OF CIVIL ENGINEERS.
THE INSTITUTION OF MECHANICAL ENGINEERS.
THE INSTITUTION OF NAVAL ARCHITECTS.
THE IRON AND STEEL INSTITUTE.
THE INSTITUTION OF ELECTRICAL ENGINEERS.

CONTENTS

LESLIE S. ROBERTSON, M.Inst.C.E.,
Secretary.

LONDON :
PRINTED AND PUBLISHED BY WILLIAM CLOWES & SONS, LIMITED,
23, COCKSPUR STREET, CHARING CROSS, S.W. ;
AND TO BE PURCHASED FROM ANY BOOKSELLER, OR DIRECT FROM THE OFFICES OF THE COMMITTEE,
28, VICTORIA STREET, WESTMINSTER, S.W.

February, 1903.

Price 1/- net.

The first British Standard, for rolled steel sections, published
in 1903, achieved a 30 per cent reduction in sizes in common
use (see page 9)

To face p. 2

The Proclamation granting a coat of arms to BSI in 1951

Board. (See Plate facing p. 90.) The membership of the Executive Board includes representatives of the Founder Institutions, the Confederation of British Industry, the leading professional, commercial and user organizations, the Trades Union Congress, the nationalized industries and Government departments. The main core of Board members are elected from Building, Chemical, Engineering, Textile and Codes of Practice Divisional Councils, and there is also a representative of the senior policy body concerned with certification and approvals, the Quality Assurance Council.

Under the five divisional councils, some 80 Industry Standards Committees and Codes of Practice Committees play a key role in BSI's activities. These committees have the responsibility for authorizing new work, allocating priorities, reviewing progress on the thousands of projects currently in hand, and approving for publication the final drafts of standards and codes of practice.

The technical committees, under which all of the technical work is carried out, are the backbone of the Institution. They number some 4500 at present.

What is a British Standard ?

A British Standard is a precise and authoritative statement of the criteria necessary to ensure that a material, product or procedure is fit for the purpose for which it is intended. Broadly speaking, British Standards fall into five main groups: glossaries, containing precise definitions of the terminology used in a particular field of technology; dimensional standards; performance standards; standard methods of test, describing how, for example, a given level of performance or a particular composition may be determined; and codes of practice.

A single standard may embody any or all of the first four features. The preparation of standard glossaries of terms and symbols is a prerequisite to effective communication, and thus it is frequently the first step towards standardization in an industry and also in international harmonization. For example, the British Standards for SI units, graphical symbols in power, telecommunications and electronics, and nuclear science terminology, are all key reference documents in their fields. British Standard definitions are the basis of sound trading in that they are recognized as the distillation of good commercial practice.

Dimensional standards ensure that the same products will be identical in shape and size within specified tolerances. They give interchangeability and should provide a rational range of sizes, eliminating unnecessary variety, and facilitating interchangeability, repairs, simplified stock-keeping, better control of production and purchasing, and, if properly used, enormous savings in time and money in design work. Typical examples of such standards are the sizes of screw fasteners.

3

Performance or ' quality ' specifications define adequate standards of performance leaving the manufacturer free to design a product that is suitable for the duty for which it is intended, and will be satisfactory and reliable in service. Specifications defining safety requirements for industrial and consumer goods (for example, domestic electrical appliances, protective clothing and life-saving equipment) form an important group in this category. The use of this type of specification is, however, dependent upon the existence of appropriate test methods. Its great advantage lies in flexibility to accommodate technological advances.

Parallel with performance specifications are standard methods of test. These make it possible to compare, under controlled and uniform conditions, materials and products intended for the same purpose. These standards greatly help the work of test houses, who know that they will be testing a material in exactly the same way as another test house, and also make possible the implementation of quality assurance schemes which are of real benefit in terms of efficient production control to manufacturers, and to purchasers, who know what they are getting. Standard methods of test also make it easier for manufacturers, when tendering, to calculate the cost of ensuring compliance with standards.

Codes of Practice set out the recommended practices for the design, installation, and maintenance of equipment, buildings and services, and are frequently quoted in legislation and local bye-laws, especially where safety is at issue. They cover, for example, every aspect of building from the basic construction and civil engineering services to ventilation, central heating, lifts and refuse disposal.

Reflections of the national scene

BSI is now geared to an output of between six and seven hundred publications every year. In the last twenty years, the average number of standards published annually by BSI has grown from 200 to 600, the budget has increased from £$\frac{1}{4}$ million to nearly £3 million; and the number of subscribing members has risen from 6300 to 14 600.

The story of the British Standards Institution is bound to reflect the larger national and international scene, as the following chapters clearly bring out. In the early 1900s standards introduced a much needed discipline into industrial practices. In two world wars, standards were pressed into national service to help speed the flow of munitions. In the austerity days of the late 1940s and early 1950s standards helped to make the most of scarce materials and goods. Standards have contributed in no small way towards the emergence of the affluent society by facilitating the mass-production of inexpensive and reliable versions of what once were ' specials ' or luxury items. Today, standards have a special significance in the removal of non-tariff barriers to trade, of vital importance to Britain's export drive.

In a narrower sense, too, standards have mirrored both the march of industry and the problems that this has sometimes brought in its wake. Thus, as the years have unrolled, BSI has set up new standards-making machinery to meet the needs of new industries, and the use of new materials and techniques: telecommunications and electronics, automobiles, refrigeration, data processing, nuclear energy, medical equipment, industrialized building, materials handling, and a host of other subjects. Parallel to these advances have come standards related to the environment, to deal with the problems of pollution, noise and waste.

The major developments of the last twenty years are treated both historically and thematically in the following chapters. First and foremost among these are the great advances in the international alignment of standards in the interests of international trade. Linked with this has been the formation of a policy to introduce and facilitate British industry's change to the metric system and the establishment of the Technical Help to Exporters Service. Among other important trends has been the development of standards engineering as a way to make optimum use of standards in industry and commerce; the development of consumer and general community representation in standards work, accompanied by the creation of a large body of standards for consumer goods and the progress of quality assurance schemes and associated test facilities, including the development of the Hemel Hempstead Centre.

These trends in particular have brought home to most people a realization of the vital part played by standards both in the smooth running of industry and in their own daily affairs. The British Standard Kitemark and other BSI marks of approval are on products in nearly every household, and many more products are designed in compliance with British Standards, though the user may not always be aware of this.

Standards and change

As a corollary to the rapid development that the past has seen, so too the standards scene is one that is constantly changing. One of the greatest challenges at present facing BSI is that of ensuring a right balance between the object of achieving uniformity of good practice and that of making provision for a rapidly advancing technology. This is an ever-present problem, but one that is now more than ever acute. Out-of-date standards are worse than useless, not least because they will tend to inhibit good design practice.

Conversely, standards that incorporate the fruits of the latest research in what may be a narrow field of technology frequently themselves provide the principal means of communicating this information to industry at large. It is therefore essential that British Standards should be regularly reviewed, and revised to take account of any technical improvements. The timing of this work must be carefully judged, and once begun it must be speedily carried out. It is also imperative that the quality of membership of technical committees should be of the highest order, containing the widest possible spread of expertise, so that new standards represent the best possible solution that existing technology permits.

There is no simple answer to the perennial problem of how to reconcile the needs of technological advance with those of national and international harmonization. Some of the steps which BSI is currently taking to ensure an optimum solution are described in the final chapter. But the basic principle that is being followed throughout is one of evolution. It is a major aim of the following account of the development and diversification of BSI's activities to provide the key to an increased understanding of this evolution and the course that this is now taking.

1

The start of it all : Whitworth, Skelton, Barry and the years to 1914

Industrial standards could well be said to owe their origin to the coming of the railways. Before the great railway boom of the middle 1800s, markets were local and what was required could be supplied from local resources. But from 1850 onwards, with hitherto undreamed of facilities for transporting goods from one part of the country to another, the world had entered a new era; the Industrial Revolution was able to move into top gear. Now the engineering shops of Birmingham, the steel mills of Sheffield, the cotton looms of Manchester had all Britain on their doorsteps – and beyond England there were further markets to conquer in all the other countries of Europe which, with England, were thrusting forward with their own railway networks and industrial development.

The vast increase in the number of production points soon gave rise to problems. First, the very diversity of sizes and qualities of similar products made in different places made it difficult for buyers to know just what they could expect to be getting, and even more difficult to re-order with any degree of assurance. Then there was the even greater problem of matching components and semi-finished products from different works together. Purchasers found that materials and components bought for the same purpose from various factories could not without modification be used together in the same ultimate product. The engineering industries in particular, requiring metals in a number of forms for conversion into rails, beams or machines, discovered that these forms differed widely in dimensions from one supplier to another and were not readily adaptable in the same structure.

Sir Joseph Whitworth was making history when in 1880 he pointed out the need for standard-sized candle butts to fit into standard-sized candlesticks! Forty years earlier, seeing the rather more serious need for standardization of the nuts and screws which were going to be used to fasten together every product of mechanical engineering, Whitworth had devised the standard screw thread that still bears his name today. It is one of the not inconsiderable tragedies of industrial history that in 1864, 25 years after Whitworth standardized screw threads in Britain, Sellers standardized screw threads in America to a different pattern. (At that time and for long afterwards the Whitworth standard thread was extensively used throughout the metric countries of Europe and indeed it was only in 1938 that Hitler banned its use in Germany.)

During the last decade of the nineteenth century technical advances in the production and use of iron and steel were running full spate. Mild-steel and wrought-iron sections were being used in bridges, ships and buildings in place of cast-iron girders. Every architect and builder ordered the particular sizes of steel sections he needed for his own work.

By 1898 steel girders were such common articles of commerce that every iron merchant carried stocks; rent and interest charges on the cost of selling them became heavy and the economics of marketing and storing had to be seriously considered.

The problem had been recognized in January 1895 when a letter in *The Times* quoted the case of a contractor who, having placed an order for girders, found that after many transfers of the order from one maker to another, they were finally supplied from Belgium. This letter produced an answer from a City of London iron merchant named Skelton who wrote:

' If the members of the engineering professions had any real interest in the English working man or showed anything like scientific method in the practice of their profession, such an instance as the one he relates would be impossible. Rolled steel girders are imported into Britain from Belgium and Germany because we have too much individualism in this country, where collective action would be economically advantageous. As a result, architects and engineers specify such unnecessary diverse types of sectional material for given work that anything like economical and continuous manufacture becomes impossible . . . no two professional men are agreed upon the size and weight of girder to employ for given work and the British manufacturer is everlastingly changing his rolls or appliances, at greatly increased cost, to meet the irregular unscientific requirements of professional architects and engineers. '

Strong words, but it was another five years before Mr. Skelton was invited to air his views at a meeting of the British Iron Trade Federation. His paper advocating standardization of manufactured steel was specially well received by a leading consulting engineer – Sir John Wolfe Barry, builder of bridges, railways and docks, and a member of the Council of the Institution of Civil Engineers. In January 1901, Sir John persuaded the Institution to appoint a committee of leading civil engineers to consider the advisability of standardizing various kinds of iron and steel sections. The Institution of Mechanical Engineers, the Institution of Naval Architects and the Iron and Steel Institute were also invited to nominate representatives and thus there came into being on April 26 1901 the original Engineering Standards Committee – forerunner of the British Standards Institution.

To its original terms of reference – standardization of iron and steel sections – the Committee shortly added the standardization of locomotives and tests for engineering materials and in 1902 with representation from the Institution of Electrical Engineers, it further widened its scope to include standards for electrical equipment.

A major achievement of early standards work was the
agreement of standards for rails (see page 9)
(*Photo courtesy Radio Times Hulton Picture Library*)

[*To face p.* 8

Standardization of tramways rails in 1903 led to a reduction
in sizes used from 75 to 5 (see page 9)
(*Photo courtesy Radio Times Hulton Picture Library*)

Sectional committees of the Engineering Standards Committee, which included representatives of industrial trade associations, technical institutions and government departments were soon at work drawing up specifications for steel sections used in bridges and building construction, railway rolling stock underframes and rails, and shipbuilding components. These committees achieved some remarkable results. The first was a reduction in the number of structural steel sections in common use from 175 to 113; tramway rails were slashed from 75 to 5. It was calculated at the time that this led to economies in production costs of something like £1 million a year. It also greatly reduced steel merchants' costs by cutting down the varieties they had to stock. Standardization actively promoted the wider use of steel partly because prices were cut and also because contractors found, for the first time, that sections ordered to the same specification were interchangeable regardless of the works from which they came.

In 1902 a deputation from the Committee met Mr Arthur Balfour, who was shortly to become Prime Minister, suggesting that the Government should make a grant for its work in support of the contribution by the engineering institutions, which was rapidly being outpaced by the demand for more standards. The first government grant in respect of the year 1903–1904 was £3000. Thus, by 1903, the foundations were laid for the world's first national standards organization – a voluntary body, formed and maintained by industry, approved and supported by the government, for the preparation and publication of technical standards that would assist Britain's industrial well-being.

At a meeting of the British Association for the Advancement of Science in August 1906 Sir John Wolfe Barry, by then chairman of the Committee, noted that British Standards had been prepared for an impressive variety of goods ranging from Portland cement to cast iron pipes, from locomotives to electric cables. 'Standardization' said Sir John, 'is a rather barbarous word but perhaps describes by a short cut and better than any periphrasis, an important departure which is of great interest to us all, the study and recognition of standard forms and qualities.' He went on, 'If overseas governments will insist upon all materials bought by them being to British Standards, they will know that they are receiving good designs and high-class materials' a thought that was to be echoed down the years subsequently and certainly one of the most dominant of BSI themes in more recent years.

Later, in May 1911, at a banquet given in his honour, Barry noted another feature which then, as now, was a cornerstone of standards philosophy: 'The organization of which we are brethren is not one for stereotyping design or any particular form of material; we have arranged that our committees sit permanently for the purpose of considering improvements or developments, advances in science or new requirements.'

By 1914 the position of the Engineering Standards Committee was assured. The Admiralty had adopted British Standard steel specifications; the Board of Trade rules for passenger steamships took note of the standards for steel forgings and castings; Lloyd's Register of Shipping had adopted British Standard steel sections for marine work; the standard for cement was widely used. Returns from the biggest steelworks showed that 95 per cent of their output of rolled sections in the year 1913–1914 conformed to British Standards and it was estimated that the whole of the material used in the construction of railway rolling stock for India was to British Standards. In 1914, with 64 reports and specifications published, there came the outbreak of war and with it a quickening in the tempo of standards activity.

2

Standards through war and peace...
and war again: 1914-1946

Throughout the First World War the Engineering Standards Committee was active in promulgating standards for the materials and munitions of war. In 1915 the Committee acquired as its secretary, Mr Charles le Maistre, who was to become a dominant figure on the standards scene over the next 30 years.

Certainly the Committee's most important contribution to the war was in laying down standards for the materials used in the newest of the fighting arms – the aeroplane. In the summer of 1917 – a black time for the Allied Powers – a new sectional committee was charged with the urgent task of setting standards for aircraft, and aircraft-engine, components. Later that year the Government set up the Air Ministry and it was decided as a priority measure to put into the Engineering Standards Committee's hands the standardization of all aircraft materials for the services. With the rising significance of aircraft as military weapons and the entry of the United States into the war, this aspect of standardization became increasingly important. In February 1918 a commission from the American Aircraft Board came to London for an inter-allied conference organized by the Committee.

The young Winston Churchill, then Minister of Munitions, opening the conference, said: ' The value of the work of the Engineering Standards Committee in other fields is well known . . . The detailed standardization of aircraft materials among the allied powers now fighting the Germans is based on principles so obvious that they really do not at this time of day require even to be emphasized.' The American delegates were given full information as to the position regarding the standardization of aircraft materials which assisted them on their return to accelerate the production of aircraft parts.

In 1918 the Committee changed its name, being incorporated as the British Engineering Standards Association on 23 May. One of the reasons for this move was that it facilitated the use of the British Standards mark, which could be applied by manufacturers as denoting that goods on which it appeared were in accordance with the relevant standard (see Chapter 12).

The uneasy peace

By 1920 there were 300 committees engaged on standardization work compared with only 60 before the war and the sale of standards had risen tenfold. During the 1920s the standards concept was developing widely throughout the British Empire (see Chapter 8) and interest was growing in other countries. In 1920 the secretary of the American Standards Engineering Committee had visited London to examine the workings of BESA.

In 1929, against a background of economic slump, BESA's work was strongly acclaimed by the Balfour Committee on Industry and Trade, which noted the general appreciation in industry of the Association's work, in which 500 committees were now engaged. The committee advocated that the Government's nominal annual contribution of a hundred guineas, far less even than in 1903, should be stepped up to £10 000 so as to lessen the Association's almost exclusive reliance for cash on industrial subscriptions. The government grant was increased but only to £3000, for the years 1930 to 1935.

On 22 April 1929, the Association was granted a Royal Charter which defined its objects and purposes:

> To co-ordinate the efforts of producers and users for the improvement, standardization and simplification of engineering and industrial materials so as to simplify production and distribution, and to eliminate the national waste of time and material involved in the production of an unnecessary variety of patterns and sizes of articles for one and the same purpose.

> To set up standards of quality and dimensions, and prepare and promote the general adoption of British Standard specifications and schedules in connection therewith and from time to time to revise, alter and amend such specifications and schedules as experience and circumstances may require.

> To register in the name of the Association, marks of all descriptions, and to prove and affix or license the affixing of such marks or other proof, letter, name, description or device.

> To take such action as may appear desirable or necessary to protect the objects or interests of the Association.

Thus, was spelled out the future working of Britain's standardizing body, and thus it has remained virtually ever since.

Up to 1930 the work had been confined almost entirely to engineering. Now, following proposals from the Association of British Chemical Manufacturers, it was decided to integrate chemical standards into the existing BESA programme. To allow for this to happen, King George V was petitioned for a supplemental charter by which the Association's name was changed to its present form, the British Standards Institution, with a General Council and three Divisional Councils to organize work in the fields of chemicals, building and engineering.

Development in the 'thirties in certification mark schemes stimulated an interest, too, in sampling methods. A theory of coal sampling worked out by two leading physicists, involving tests on samples from over 100 000 wagons of coal, resulted in a series of British Standards for the sampling of coal and coke. Sampling methods were also devised for testing such products as

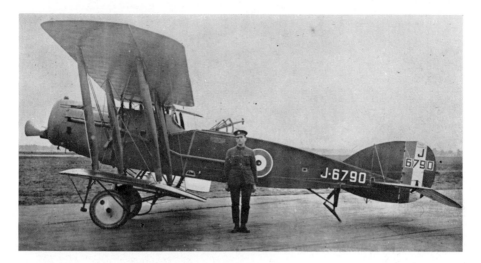

During the First World War production of aircraft acquired
a new importance. In 1918, the Engineering Standards
Committee took over the responsibility for aircraft specifications (see page 11)
(*Photo courtesy Imperial War Museum*)

The first Secretary of the Standards Committee, Leslie
Robertson, was drowned when HMS *Hampshire* (pictured
below) was sunk in 1916 on the way to Russia. Lord
Kitchener was also among those who lost their lives
(*Photo courtesy Imperial War Museum*)

[*To face p.* 12

The standing type of 600 British Standards was destroyed
when Waterlow's printing works were bombed in 1940. A
new style was adopted when the standards were reset later
(*Photo courtesy Waterlow and Sons Ltd*)

gas mantles and electric lamps and these methods were later to play an important part in developing the techniques of quality control.

Another important development between the wars was the emergence of ' codes of practice '. By 1939, over 850 British Standards for a wide range of engineering, chemical and building materials were providing the technical basis for industrial contracts running annually into millions of pounds. It was becoming increasingly clear that fitness for purpose of a material, article or appliance depended also upon correct installation, maintenance and use. Sometimes requirements covering these features were included in a standard – as in BS 449 first issued in 1932 for structural steel in building. Such standards became known as codes of practice and the term was also applied to standards dealing with testing procedures such as acceptance tests for steam turbines. In 1939 a Codes of Practice Committee for the building industry was set up, but was not called together before war broke out.

In 1928 there was formed the International Standards Association. At that time, UK industry had decided not to support BSI's participation in that work. It was not until 1938 that there was a change of attitude, but by then the clouds of war were again gathering.

Emergency standards for war

Immediately war broke out in September 1939 ordinary work was put on one side and efforts were concentrated on producing ' war emergency standards ' as required by government departments. Much of the normal committee procedure was by-passed, what committee meetings there were often being held in centres nearer manufacturing areas. Standards, it soon became clear, had a much larger part to play in the manufacture of weapons and production of consumer goods than was even the case in the First World War: in keeping to a minimum the number of types, sizes, grades and component parts and in ensuring the greatest output from the labour and resources available.

The first emergency work was on behalf of the Ministry of Home Security. Nearly 70 standards were issued for such items as black-out material for windows, stirrup-pumps with which civil defence workers could tackle fire bombs, and components for the construction of back-garden air-raid shelters. Another important series of standards were those for special lighting fittings such as those used for the signs on air-raid shelters. Extensive experiments in blacked-out London streets and other big cities resulted in the fixing of a permitted level of street lighting. In charge of all this activity was Mr Percy Good who in 1941 had become director of BSI.

For the Ministry of Production BSI became deeply involved in packaging matters. From the lavishness of pre-war packs, Britain had to turn to the barest minimum in the way of outer wrappings for goods. An investigation into the most economic packaging materials, minimum practicable range of sizes and whether or not pre-packing was necessary, led to a series of standards which cut the amounts of paper and board and tinplate in use by over half. In 1943 BSI issued its 'packaging code' to rectify the serious situation in which vital supplies for the fighting services overseas were arriving in unusable condition because of unsuitable packaging. The code has been revised over the years and remains today as a guide to industry on the best choice of packaging materials for any particular job. From the needs of these times of crisis has grown up an extensive programme of standards for packaging and transport, ranging from packaging for medicines to the transport of animals by air.

Further work on quality control methods was done for the Ministry of Production – the object being to quantify on a statistical basis the routine sampling and inspection of raw materials and finished goods, so as to reduce as far as possible any waste due to bad workmanship. Another rather odd job which BSI agreed to take on in the absence of any other suitable body was the preparation of guidance for industry on such matters as production control, office organization, pay-roll methods, stock control and storekeeping. (This work was passed in 1947 to the newly-formed British Institute of Management.) Here was an early example of BSI being asked to diversify into quite new fields of activity. There were to be many further requests in the years following.

For another government department, the Ministry of Supply, BSI helped to co-ordinate the steel specifications of the UK, Canada and United States, and to assist the comparison of materials supplied by the three countries. All told, BSI produced over 400 war emergency standards between 1939 and 1945. Among the savings which resulted, it was estimated that the emergency standard for tins and cans saved 40 000 tons of steel in its first year. The publication of another standard enabled one firm making clockwork mechanisms to reduce from 600 to 20 the number of teeth-forming tools.

Another important departure was the move by BSI into consumer-goods industries. In 1940 a BSI committee was established to help the Board of Trade in laying down standards for the 'Utility' clothing, household textiles and furniture which became so much a part of Britain's austerity way of life. Here was the start of BSI's growing involvement in standards for the ordinary shopper (Chapter 11). Again looking to the future, BSI committees were also making recommendations, at the request of the Ministry of Works, as to the building standards which would be required for housing development after the war.

Towards the end of the war it was decided that BSI should have a president as well as a chairman of its General Council and in 1944 Lord Woolton was elected as first president of the Institution. The popular eminence of Lord Woolton as a leading figure of the Government of the day brought to the office of president the symbolic recognition of the role of BSI in national life, beyond the immediate context of technical standardization.

3

Standards for a post-war world: the Lemon and Cunliffe Reports chart the way ahead

When peace came in 1945 it found a Britain drained as never before of strength and energy, its industrial resources at a very low ebb, large areas of cities devastated by bombs, short of raw materials, of machinery, of man-power. All had been poured into the effort of winning the war. Now, there was the formidable task of rebuilding and renewing – homes, factories, communications. In this gigantic job of re-building, standards were as never before to prove their worth in securing the optimum use of precious raw materials, eliminating wasteful diversity, generally making the most of scarce resources of man-power. Simplification through standardization was made necessary by shortages – although it became increasingly useful as a tool for boosting production.

Standards were also to be called upon to help industry sell more abroad – the export drive was to be the main plank of government industrial policy through successive post-war administrations. Certainly in the late 1940s there was a pressing need to earn our way in the world through re-establish-ment of our export trade and as living standards rose and the bill for im-ported goods has gone up, so the need to increase earnings from overseas trade has also increased. The war was therefore not long over before BSI's industry standards committees were being reminded that they should study the standards they were revising and preparing with a special eye on their application to the sale of goods overseas, to make sure that the goods concerned were going to comply with the requirements of the countries for which they were intended. In 1946 the setting up of the International Organization for Standardization pointed to the desire of all the major industrial countries for collaboration between them to this end (Chapter 9). In this work and that of the International Electrotechnical Commission just resuming after its war-time break, BSI was playing a very active part, or-ganizing many of the international meetings in London and taking over secretariat responsibilities for many of the new committees being set up – steel, aircraft, textiles, rubber, pallets, steam turbines, electric cables, to name just a few. The first in a series of British Commonwealth standards conferences was held in London in 1946.

European collaboration on standards began in the immediate post-war period with the setting up by the Organization for European Economic

Co-operation of a special advisory group for increasing productivity through standardization, of which BSI was a member. Efforts towards simplification in the industries of West European countries were being undertaken under the banner of the European Productivity Agency (Chapter 10).

These years after the war were also to see a great onrush in technological advance with the birth of new industries, new techniques, new materials. Space travel was still a dozen years away but already the developments in electronics and automatic data processing, the introduction of new metals and plastics which were to make it possible were under way. In these rapidly advancing fields of technology, standards were also required to set the right pattern for future orderly development.

On a more humdrum level, BSI was now taking up programmes of work which for six years had remained in abeyance. The standards produced to deal with the emergencies of war were being brought under review to decide whether they should be withdrawn, continued, or modified for peace-time use. Above all, consideration had urgently to be given to the needs of every industry for a host of new standards to meet the requirements of a new era. Reflecting this growth in work, from 1945 to 1950 the number of industry standards committees rose from 40 to 60, and the number of firms subscribing to BSI went up by more than 25 per cent.

Work on textile standards began during the war, continued and grew and to meet the new demands, a Textile Divisional Council was set up in 1947 to join the existing ones for building, chemicals and engineering. A branch office was opened in Manchester to serve as a centre for standards work on textiles and textile machinery.

The construction industries, faced with their large programme of reconstruction, were in need of codes of practice to guide them in their work. The Ministry of Works accordingly set up a new council to deal with this subject. BSI, which had set up its own codes of practice committees in 1939, was invited to collaborate and to be responsible for circulating the draft codes and publishing them as British Standard codes when finally approved.

Attention was also given after the war to the special requirements of local authorities and hospitals whose annual purchases of equipment ran into many millions of pounds, and who tended each to lay down their own private purchasing specifications. Two BSI advisory committees – one dealing with local authority, the other with hospital, purchasing – were set up to examine the suitability of ordinary existing standards and see what new ones were required, and to draw up common programmes to which all local authority and hospital purchasing officers could work. Another important advisory committee was set up in 1946. Representing the main Scottish professional, municipal and government organizations concerned with building its purpose is to ensure that British Standards take account of Scottish practices.

The period of the late 1940s was not only one of shortages, of tightening the national belt, of making-do-and-mend. It was also a time of re-appraisal; amid the hurry and scurry of trying to get the wheels turning again, BSI and its supporters in industry and government were giving thought as to the future direction and organization of standards affairs in Britain.

The Socialist government of the day, with its commitment to industrial planning, had set up working parties to consider the problems of major industries. The reports of many included consideration of the value of standardization. Arising from this, there was a general welcome for the decision by the Ministry of Supply in 1948 to set up a departmental committee under Sir Ernest Lemon, a leading engineer: ' to investigate the methods by which manufacturers and users of engineering products determine whether any reduction in the variety of products manufactured is desirable; to report whether these methods are adequate and what, if any, further measures should be taken by industry or government to ensure that such simplifications as are determined are put into effect.' The Committee's findings, with possible advocacy of some change from the voluntary system of standardization, were to have powerful repercussions on the future of BSI.

At the centre of the Committee's recommendations when they were issued a year later was the call for more specialization, simplification and standardization while avoiding any harmful rigidity in technique. Changes in production methods following the discoveries of science and development of mechanical processes and automatic controls, demanded a reduction of unnecessary and haphazard variety of output.

Particular attention was given to whether maintenance of the voluntary principle for evolution and acceptance of standards was likely to provide the best results. The Committee said it was convinced that the wider application of specialization, simplification and standardization would best be achieved by voluntary action. The report said: ' We appreciate the great value of the work which BSI has carried out in the past and which has been on an ever-increasing scale. Indeed the evidence which we have examined shows that this country has in the past been second to none in the preparation of national standards. We are agreed that BSI is the appropriate body to co-ordinate the views of the various interests concerned in drawing up standards on a national basis but we believe that its staff and facilities must be strengthened and extended if it is to play its proper part in the work which now has to be handled. '

The Lemon Committee put forward 20 recommendations aimed at reducing wasteful variety of output both in individual firms and in complete sections of industry. The recommendations set the pattern for standards development in the years ahead – including the more concerted representation of user organizations in standards-making, the far wider employment of ' standards engineers ' in individual firms, encouragement of large purchasers (e.g. government departments) to buy to British Standards, creation of standards ' handbooks ' covering all the standards in an industry, and the strengthening of BSI facilities so as to speed and intensify its work.

In this final connection, the Committee noted that BSI would need to have more money and that while industry should be ready to bear a proportion of the extra cost, it was essential that the Government contribution should also be increased.

The Government accepted the broad conclusions of the report – that standardization was an activity for which industry itself must continue to be responsible, and that standards should not normally be legislatively imposed, but that the progress of the standards movement should be encouraged by administrative action. From a figure of £44 754 in 1948–9, the Government had by 1950–1 stepped up its grant-in-aid to £90 000. Before this occurred, the President of the Board of Trade decided to appoint another departmental committee ' to consider the organization and constitution of the BSI, including its finance, in the light of the increasing importance of standardization and the extended size and volume of work likely to fall on BSI in the future '. Mr Geoffrey Cunliffe, later to hold high office in BSI, was appointed chairman.

The Cunliffe Committee reported early in 1951 and, like the Lemon Committee before it, foresaw no need for any major departures in the workings of BSI. The machinery was ' well adapted for carrying out its present work and the increased work which may be expected to fall on it '.

The Committee advised that subscriptions from members should be increased and that wider industrial membership should be enlisted. It also recommended that the Government contribution should be a minimum of £90 000 and that for the future it should be equal to the income from industrial subscriptions. There were also a number of other recommendations which were greatly to influence the course of action in the following years: a general increase in user representation on technical committees; representation of domestic users – almost the beginning of BSI's important role in consumer standards and consumer protection – and the use in this connection of the women's organizations, leading to the formation of the Women's Advisory Committee (Chapter 11); encouragement of the wider use of certification marking; the adoption of a more positive policy on publicity – BSI ' should take such steps as may be necessary to make the theory and practice of standardization better known and understood both by industry and the general public '.

On 6 December 1950, Mr Percy Good who had been associated with BSI for 37 years, and been director for nine years, died and Mr H A R Binney, who had joined the Institution only two months before as director-designate, was appointed to succeed him – the start of a new era in BSI's affairs, one in which the Institution's prestige and influence at home and in the international field were to grow immeasurably.

Mr L G Watkins, organizer of the 1951 Golden Jubilee
Standards Exhibition, points out an exhibit to Lord
Monckton (see page 19)

[*To face p.* 18

Delegates to the 2nd Commonwealth Standards Conference, held in London in 1951, on a visit to Hampton Court (see page 19)

4

The 1950s: focus on expansion

Between 1950 and 1960, more standards were produced than in the entire previous fifty years of BSI's history. The scene for this remarkable expansion was aptly set, in June 1951, with the celebration of BSI's golden jubilee.

Representatives from the standards organizations of 25 other countries – a measure of the way the standards movement had grown since its inception in London in 1901 – came to BSI to participate in these celebrations and to attend the Second Commonwealth Standards Conference. Among the highlights of the jubilee – the receptions, the special excursions for the oversea visitors, a banquet given by the Lord Mayor at Guildhall, a visit to the Festival of Britain on the newly-developed site on the South Bank of the Thames – probably the most outstanding was the world's first major standards exhibition. Held at the Science Museum in South Kensington, it was entitled ' British Standards – the Measure of Industrial Progress '. Fifty-four industries co-operated to show the advantages of using standards.

Appropriately, too, the jubilee celebration was the ocasion for the publication of the first history of BSI – ' Fifty Years of British Standards '.

The following month Sir Hartley Shawcross, then President of the Board of Trade, announced in the Commons the Government's broad acceptance of the Cunliffe Committee recommendations, and the way ahead for the next two decades had been mapped out.

Dimensional co-ordination — a new approach in building

The early 1950s was a time when the necessary controls of war and the immediate post-war years could at last be thrown off. Money and other resources were becoming a little easier; people were demanding some improvement in living standards; above all they wanted homes. This was the period of Harold MacMillan's housing drive with the call for 300 000 houses a year. For BSI it meant a crash programme of standards to achieve simplification and economy in use of still scarce materials, mass-produced joinery, heating and cooking appliances, the introduction of substitutes such as plastics for metal fittings.

New techniques to speed building were being introduced. A government committee looking into ways for the quicker completion of house interiors recommended that one technique worthy of study was that of dimensional co-ordination. Dimensions of the 50 000 different building components in use at that time, although some 18 000 of them complied with British Standards, nevertheless showed great variation. The production and assembly on-site of different products coming from factories all over the country was often complicated by the fact that each producer had tended to work to sizes and forms which were either traditional or special to the firm. The increased extent to which building components were being produced by industrial, mechanized, methods had made a solution to the problem much more urgent. Dimensional co-ordination provides a method whereby the sizes of all components may be co-ordinated one with another, using dimensions which are based on multiples of a small unit size to which the name ' module ' has been given.

From 1953 various BSI building committees were looking into the possibilities and it was agreed that a scheme of investigation should be put in hand, in which UK activity would be integrated with European studies under a project sponsored by the European Productivity Agency, for which BSI provided the secretariat under the leadership of Mr Bruce Martin. The EPA's first report was published in 1956. Second phase of the operation was to convert theory into practice, translating the information contained in the report into ordinary building practice, and constructing buildings to modular precepts; one of the UK contributions was to be the first stage of the BSI test centre at Hemel Hempstead (Chapter 13).

Speaking in December 1956 to the Modular Society, Mr H A R Binney drew an analogy between the changes then taking place in the building industry and the revolution which had altered the face of the textile industry in the eighteenth century. ' With the spinning jenny and the loom, the textile industry moved out of the cottage into the factory, ' he said. ' The average builder in this country is still concerned with constructional work on the site, but the growth of industrialization is already turning the builder's task into one of assembly rather than construction . . . the chemical and engineering industries are now producing new building materials and components. '

Move to British Standards House

In August 1953 BSI was able, after a 52-year sojourn in its Victoria Street premises, (described in the Cunliffe Report as a ' rabbit warren of rooms ') and various nearby adjuncts, to move into a new home – British Standards House, 2 Park Street, W.1 – which appeared at that time, with 45 000 sq ft, to offer all the space that would for a good many years be sufficient to produce standards effectively and efficiently. That within a comparatively short space of time extra accommodation had to be found is a measure of the unprecedented rapid rate of expansion that was to occur in the 1950s and 1960s. But 2 Park Street has remained the hub of Britain's standards world.

Another landmark in BSI's history came in 1954. Following detailed negotiations, endorsement was given by the Government through the Minister of Works to an agreement, reached between BSI and the fourteen major professional institutions concerned with building and engineering, for the creation of new BSI machinery for the full preparation and publication of all engineering and building codes of practice. Thus was born the Codes of Practice Council of BSI, with all the sectional work stemming from this development.

Industry was constantly seeking more efficient ways of using standards. Work was in hand on the co-ordination of specifications to ensure that where they were common to a number of different industries the most economic range of products suitable to all the industries would be forthcoming from them. In 1954 work was begun, for example, to co-ordinate standards of common interest to mechanical and chemical engineering, and the petroleum equipment industries; thus in the field of pressure vessels, when BS 1500 came to be revised, its provisions were each generally applicable to all three industries.

Throughout industry as a whole there was at this time mounting interest in management techniques. Encouraged by the Anglo-American Council on Productivity, companies of all kinds – long insulated from the rough and tumble of tough competition – were seeking to learn from their American counterparts methods they could adopt in boosting their own outdated approaches. A cornerstone of the Productivity Council's gospel, to be repeated during the 1950s at conferences throughout the UK, was that industrial efficiency and output could be greatly increased if only more firms would apply the techniques of simplification, standardization and specialization. The 'three S's' became watchwords for a decade (Chapter 6).

Standards engineering

BSI was increasingly coming to accept that in addition to preparing and publishing standards, industry was looking to it to fulfil yet another task – namely help in applying those standards on the factory floor. Acceptance of this additional responsibility was encouraged by the emergence of a new individual in industrial management – the standards engineer. In May 1955 came an event of considerable significance, the first in what was to become a series of annual meetings of engineers and others responsible for applying standards in their various organizations. This first modest gathering, organized jointly by BSI and the Institution of Production Engineers, took place at British Standards House and attracted an attendance of some 100 people (Chapter 6).

One of the results was the setting up of a panel of engineers to advise BSI on the application of standards in industry. Among suggestions made at this first standards engineers' conference was that ' data sheets ' as used in some Continental countries should be prepared, in which essential features extracted from British Standards could be summarized for easy application by smaller firms on the shop-floor. The first of these new summary sheets was published shortly afterwards.

New fields of work

Throughout the decade standards were being demanded for new fields of activity – BSI had come a long way from the narrow confines of engineering standardization of its earlier years. From the early 'fifties onwards work in the technical committees was very closely to mirror the changing industrial and social scene.

Man's increasing concern with his environment had led after much effort to the passing of the Clean Air Act, destined to cure London and other big cities of the killing ' pea-souper ' fogs which were still common for several years after the Second World War. New standards were prepared for the various items of equipment needed to give effect to the legislation – for smoke viewers and smoke density indicators to help check air pollution from boiler plant, and methods for measuring the amount of grit in smoke emission, for example.

Safety of the individual was to be a recurring theme of standards work. In 1953 had been published a key standard in this field – a specification for motor-cyclists' helmets. Three years later it was revised in the light of investigations carried out by the Road Research Laboratory, and this new edition was to form a basis for international standardization. In 1956 was issued another key standard in the safety field – a specification for a new and virtually foolproof type of fireguard for open fires.

This year also saw the beginning of work on a programme of standards for the developing science of nuclear energy. The range of standards work was becoming increasingly varied: recommendations for pictorial marks to go on packages to reduce risk of breakages; a system of safety colours for use in industry (including the use of orange and black stripes to warn of hazards); a glossary of terms used in work study; school and office furniture designed anthropometrically to ' fit ' the user; standards for data-processing equipment, chromium plating, dividend warrants; recommendations on the carriage of various types of animals, from day-old chicks to crocodiles, in aeroplanes. If any criticism of BSI was to be made at this time, it would have been that the Institution was perhaps over-willing to respond to appeals for help. In fact, with so many worthy interests clamouring for BSI's services it would have been difficult to refuse to ' take the job on ', off-beat though some of them were.

Various other innovations were being successfully tried out. For example, the completion of a highly technical code of practice on frost precautions in buildings was brought to the attention of the man-in-the-street by means of a simply-written leaflet telling householders how to avoid winter freeze-ups; the leaflet was distributed by water authorities in millions.

In 1957 BSI pioneered the introduction of the now universally-used 'international' sizes of paper – the 'A' series. At the time there was much opposition both from the paper trades and industry as a whole to this attack on the old familiar octavo, quarto and foolscap. BSI stood firm, pointing to the benefits from simplification in documentation that would eventually result, to the fact that 'A' sizes were already in use in 26 countries and had been accepted by the Commonwealth countries meeting at the 1957 standards conference in Delhi.

Work was in hand on other fundamental and far-reaching standards matters. The 1939–1945 war-years had brought into sharp focus the difficulties which arose through lack of interchangeability between products and components coming from different factories. Engineers had come to realize that standards for components or end-products could not be effective unless basic standards governing design and measurement were first established. By now industry had got down to the task of laying down these fundamental standards on which virtually all production could be said to depend. In 1957 came a new edition of BS 308 which was effectively to unify engineering drawing practice; not surprisingly this was quickly to become a BSI best-seller with no fewer than 15 000 copies being sold in six months. There followed BS 1916 *Limits and fits* which though adapted to our inch measurements was founded on international practice. This standard alone opened immense possibilities for dimensional interchangeability of all manner of parts and manufactured articles.

International aspects loom large

It was during the 1950s that international standardization began to loom very large indeed in BSI's scheme of things – reflecting the need for a sustained increase in the volume of Britain's export trade. The increasing tempo of standards activity in other countries was a clear enough warning that we at least, with our whole future in the hands of our export salesmen, could not allow the creation of a new set of technical standards barriers to trade. The decade saw growing participation and initiatives by BSI in the technical committees of the International Organization for Standardization and International Electrotechnical Commission. Industry as a whole and BSI's committees in particular were encouraged in this by the ten-man Export Panel, set up in the summer of 1954 under the chairmanship of Mr Lincoln Steel of Imperial Chemical Industries and the chairman of the FBI's Overseas Committee (Chapter 10).

London was the venue in the summer of 1955 for the annual meeting of the IEC Council and for meetings of no fewer than 28 technical committees with interests ranging from radio-communication to switchgear, from insulators to lightning arrestors. Dr Percy Dunsheath, a leading figure in the electrical world, became the new president of the IEC at this series of meetings.

At the end of 1955 *BSI News* made its debut as an adjunct to the old 'Information sheet' which for a good many years had kept subscribers posted as to new BSI and overseas standards publications. The news pages now issued aimed at keeping members more fully informed on all aspects of standardization, with special reference to international work as it affected British industry. The pages of *BSI News* were to feature a variety of exciting developments on the ISO and IEC fronts such as the announcement in May 1956 that a 20-nation conference meeting at Southport had unanimously agreed to adopt a single system for expressing the linear density of all fibre yarns and threads. Agreement on the new unit, to be called 'Tex" was described as an 'outstanding achievement which could have an important bearing on the future of the textile industry all over the world.'

In June 1958 came the most notable of international affairs of the decade – the three-yearly assembly of the ISO at which for the first time it was Britain's and BSI's privilege to act as a host to the 40 countries then in membership of the ISO. Under the chairmanship of Sir Roger Duncalfe who had been ISO president during the past three years, over a thousand delegates descended on Harrogate from all the corners of the world with specially strong teams from the United States and Russia and from all the European countries. For BSI it was something of an organizational feat, the arrangements being in the hands of 70 members of its staff transferred to Harrogate for the fortnight of the conference. Comprehensive services for document reproduction had to be laid on; postal, banking and travel facilities were provided; a bulletin for delegates was published daily in English and French; a team of interpreters was on duty to help conduct the business sessions in ISO's three official languages, English, French and Russian.

All the 15 technical committees meeting at Harrogate were able to report successes by the time their meetings had ended. The food-containers committee had, in the words of the leader of the French delegation 'crossed the frontiers' after ten years of striving to reach agreement. The nuclear energy committee took its first important steps towards the evolution of agreed international standards. Sir Roger Duncalfe said in his opening presidential address: 'Probably we should all agree that a peaceful future for our civilization and a great measure of happiness for all mankind depend more upon a steady rate of economic growth and a continuing expansion of world trade than on any other factor. So we must break down the barriers to that expansion.' The breaking down of barriers was very much the order of the day at the Harrogate ISO meetings.

BSI has always entered wholeheartedly into relationships with the standards bodies of other countries and has been ever-willing to assist

The BSI Test Centre at Hemel Hempstead, pictured under
construction in 1959, was the first building in Britain to
be built according to modular precepts. This formed the
British contribution to the European Modular Coordination
Study (see page 20)

[To face p. 24

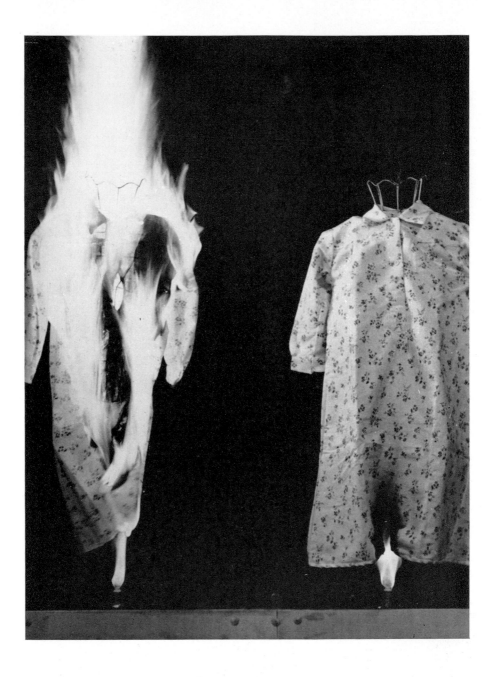

Rising concern about the flammability of fabrics led, in
1957, to the publication of a report by BSI (see page 25).
A British Standard for flame-resistant fabrics followed later

developing countries to set themselves on the right standards road. Every year BSI has welcomed to its doors those from other countries wishing to learn about the application of standards. In turn, BSI staff have visited many countries to participate in conferences and generally make their knowledge and experience more widely available. No one has contributed more to this process of imparting experience than BSI's chief for so many years, Mr Binney. In 1958, for example, he was touring Middle Eastern countries, lecturing in the Lebanon to a 12-nation conference on the role of standards in developing a national economy, and discussing standards matters with the governments of Iraq and Iran.

Consumer affairs

Around the mid-1950s the climate for further activity on the consumer-goods front was becoming extremely favourable. BSI's Women's Advisory Committee was already well-established, the number of consumer-goods standards published and under way had assumed quite respectable proportions, as had also the number of products carrying the BSI Kitemark, and there was enormous interest in the whole question of a better deal for shoppers – in Parliament, the press and broadcasting, and in public debate generally. In January 1955 BSI responded to this interest by establishing its Advisory Council on Standards for Consumer Goods which for the next few years was to make much of the running in consumer protection affairs and was to lay the foundations for a whole new set of ideas. In 1957 came the launching of the Council's revolutionary publication *Shopper's Guide* which was to pull no punches in the frank appraisal of consumer goods of all kinds (Chapter 11).

At the time the *Financial Times* aptly commented: ' The consumer today is offered so bewildering a choice of goods and so many items embodying the most advanced technical skills, that ordinary common sense and experience are of comparatively little help in distinguishing them . . . Above all there is a need for the maximum possible information to be given to the public. Precise descriptions of contents, prices, labelling of materials, precise specifications of performance – these are the ways that the manufacturer should seek to gain the consumer's confidence and allay his fears. '

One area in which concern had been mounting in the 1950s was in regard to the toll of deaths and serious injuries arising from clothing burns. Was there something the matter with clothing fabrics; could not safer ones be used? A technical committee was set up to see whether it was possible to grade fabrics according to their relative degree of flammability. The committee's report, published in May 1957, came out strongly against the idea – because most ordinary lightweight fabrics of the kind used for women's and children's clothes were found to burn at the same speed. The committee's chief recommendations were for the more extensive use of effective fireguards (like the British Standard type which could be fixed securely to a fireplace surround), and greater care with matches and flammable liquids. It was recommended that children should wear close-fitting pyjamas instead of nightdresses. A widely-distributed leaflet on the danger of clothing burns helped to focus public attention on this subject. Work was already in hand, too, on a new British Standard which was to lay down just what could be expected of any fabric to which the term ' flame-resistant ' was to be applied. Legislation, quoting the standard, was to follow and this eventually had the effect of making it an offence to sell children's nightdresses not made of a ' low-flammability ' material; women's nightdresses, if not of a low-flammability material have now to carry a label warning of the danger of fire.

In July 1959, reflecting a feeling in many quarters that a new look should be taken at the whole field of consumer protection, and as a result too of BSI representation that it could hardly take its consumer assistance activities further without a new impetus and authority, a government committee was set up under Mr J T Molony, QC, to study the question of consumer protection and to see what changes might be required in the law.

At this time, too, BSI had been taking a leading part in discussions with other organizations as to the possibility of setting up an approvals and certification scheme for domestic electrical appliances; the signs were that other countries, particularly those of western Europe, were going to have electrical approvals schemes and without something equivalent, our own industry would be at a disadvantage. In November 1959 an organizational framework was set up for the British Electrical Approvals Board for Domestic Electrical Appliances. Its board of management consisted of representatives of the electricity supply industry, equipment manufacturers and BSI, together with the electrical contracting industry, wholesalers and retailers. The object was to arrange for the testing of all types of household appliance and to publish a list of approved models. Approval was to be on the basis of British Standards and BSI undertook the quite formidable task of producing all the new and revised standards required – for everything from kettles to washing machines, from heaters to hairdriers. The mark of approval was to be a modified ' Kite '.

From BSI's point of view one of the most important happenings of 1959 was undoubtedly the opening of its own test centre at Hemel Hempstead – the start of what was to become a vital part of the standards operation. Initially used for the work of approving electrical equipment intended for export to Canada under the special agency arrangements operated by BSI for the Canadian Standards Association, the centre was also shortly involved in testing under the many important Kitemarking schemes under way: for motor-cyclists' helmets, fillings for mattresses and pillows, oil heaters and car safety belts and many others. The centre was notable for a number of unusual design features; the most important was that the building as a whole and all its component parts were based on the four-inch module advocated as an international standard and put forward as a recommendation in a draft British Standard. It was in fact the first British example of the four-inch module applied in ordinary practice.

5

The 1960s: Standards activity keeps step with growing affluence

The decade of the 1960s is still too close to us to be considered as 'history' and this chapter therefore is less concerned with detailed events than with the development and conclusion of earlier themes and the introduction of new ones. These years were to be extremely important in shaping BSI's approach to challenging new problems.

The economic background at this time was one of growing affluence. Professor A A Walters, addressing the 1963 standards conference, traced Britain's rate of economic growth from the inter-war years 'when our $2\frac{1}{2}$ per cent rate of expansion was probably better than in any other of the advanced countries', through the 1950s 'when we were doing worse than anyone else', until in the 'sixties the economy was showing 'the first slight tremors of expansion'. On the standards front it was a time of broadening activity in every area, of an ever-lengthening programme of work, the approach to which was being influenced more and more by what was going on internationally in ISO and IEC discussions and, as the decade developed, in the forum of European standardization.

Widening range of standards work

The BSI annual reports for this period offer evidence of the extensive range of subjects covered in the standards and codes under preparation – a guide to aerodrome lighting, unit heads for machine tools, pallets for materials handling, rules for the description of fabrics, measurement of vehicle noise, fire precautions in high-rise flats, lifejackets for yachtsmen. This brief list of some of the publications of 1961-2 is itself a microcosm of some of the social and economic concerns of the time: the growing traffic in civil aviation, automation in industry, concern about environmental noise, building upwards to cope with the population explosion and shortage of building land, the widening leisure pursuits of the man-in-the-street.

The growing demand for standards was reflected, too, in the number of interesting 'off-beat' standards that were being developed in the early 'sixties: for the safety of children's toys, for the performance of musical instruments played in schools (a fine opportunity, this, for the press and television to have some fun with standards!) for the more comfortable carriage of animals of various kinds in aircraft – very necessary standards in view of the appalling conditions of much of this traffic before BSI, at the request of animal welfare organizations, took a hand.

Progress in safety and consumer protection

Personal safety was very much to the fore in the 'sixties. In 1961 the Institution set up an Advisory Committee on Personal Safety to take soundings and secure advice from such authoritative sources as the British Medical Association, the Royal Society for the Prevention of Accidents and the Fire Protection Association, on the need for new or improved standards. The committee was to press hard and successfully in the following years for sound technological solutions to problems of personal safety, particularly in the field of domestic appliances.

On the national consumer front the report of the Molony Committee (see page 74) endorsed BSI's role in producing consumer standards, as well as its pioneering effort since 1955 to give direct advice and guidance to shoppers. With the forthcoming establishment of a new Government-sponsored consumer body, the responsibility for the work hitherto carried out through the BSI Consumer Advisory Council and its magazine *Shopper's Guide* was transferred elsewhere (see Chapter 11).

Standards for new technology

Many topical standards were being produced at this time for the new technologies with which BSI was becoming increasingly concerned – in the fields of electronic equipment, reprography and data processing, hovercraft and aerospace, for example.

One effect of the wider use of British Standards was an increased demand for co-ordinated and authoritative methods of test for a wide range of materials – not just traditional materials such as metals, leather and textiles, but the newer synthetic materials, such as plastics, as well. The development of standard methods of test formed a large part of BSI's activity in the early 'sixties, and much of this activity was related to corresponding international work.

Discoveries of large reserves of natural gas in the North Sea in the early 'sixties had decided the Gas Council to convert the whole country from town gas to the new fuel over the 10-year period 1966–1976. Liquefied natural gas shipped from Algeria would also be fed into the grid. To facilitate the change-over to natural gas BSI put in hand an important programme of new and revised standards including a new test schedule for gas plants, a design code for suitable carbon-steel pipelines and a series of standards and codes covering natural gas burning appliances and their installation.

Developments in building

The use of new materials, notably plastics, and the increasing production of industrialized building components, were beginning to make their impact on the construction industry.

In 1963, a two-year review of the industry's standards needs by a committee under Mr Christopher Needham, a leading municipal engineer, identified 120 products for which standards were required. The committee commented that the greatest single need of the building industry was for co-ordination of the dimensions of building products 'without which rationalization of the industry cannot be achieved'. It also stressed the need for increased financial support from the industry for national standards work. Further support for these arguments came from the Government's new Building Regulations Advisory Committee, which pointed out the importance of standards and codes as a means of implementing the 'Model' building byelaws and national building regulations to be published in 1965.

The following year, Mr Edward Heath, then Secretary of State for Industry, announced in the Commons that the Government grant to BSI was to be increased by £25 000 for 1964–1965 and by £50 000 for 1965–1966, the extra money to be devoted to work on building standards and codes of practice.

Further research on dimensional co-ordination led to the publication in 1966 of a key standard for this subject. It is interesting to record that the draft of this standard, circulated for comment in the summer of 1965, was written in metric units in anticipation of the Government decision on metrication. From now on, it became possible to combine the introduction of dimensional co-ordination with metrication (see Chapter 7).

Quality assurance

There was renewed interest towards the end of the 'sixties in the application of quality assurance schemes both nationally and in the international context. The Ministry of Technology in 1968 set up a committee under Sir Eric Mensforth to consider the future extension and co-ordination of quality assurance schemes, and the publication of the Mensforth Report in 1970 was followed up in BSI by the establishment of a Quality Assurance Council (see page 84).

International standardization dominates the scene

Overshadowing all BSI work in the 1960s was an increasing concern with international standardization. UK activity was in fact reflecting world-wide preoccupation with the subject (Chapter 9), typical of which was the interest aroused by a standards conference held in Cairo in January 1961. This was attended by 500 delegates from Arab countries, and there were speakers from BSI and thirteen other standards organizations. The emergent countries of Africa, Asia, and Latin America were now coming to realize the significance of standardization for their industrial development in the context of a world economy.

Drilling for natural gas in the North Sea. BSI put in hand
an important programme of new and revised standards to
facilitate the changeover (see page 29)

[*To face p.* 30

Dr V V Boitsov, Chairman of the USSR Committee for
Standards, Measures and Measuring Instruments, and Mr
G H Beeby, Chairman of BSI, signing the Anglo-Soviet
agreement on scientific and technical cooperation in London,
January 1968 (see page 31)

From the UK point of view BSI was especially concerned with events in Europe. The late 1950s had seen the hopes of a general European free trade agreement come to nothing, and the setting up under the Treaty of Rome of the European Economic Community. The dangers of rival standards between the Common Market 'six' and the European Free Trade Association 'seven' were all too apparent. Thanks to strenuous efforts on the part of both groups, led by the standards chiefs of the UK and Germany, a rapprochement was secured at a meeting in Zurich in June 1960 and from this date stems an increasingly fertile period of collaboration on standards matters between the countries of Western Europe, of particular significance for the later enlargement of the Common Market (Chapter 10).

Commenting on these developments in October 1966, Mr Binney said: 'Every day, pressure is being stepped up to gear the standards of British production to the needs of the export trade. "Going metric" is one aspect but by no means the whole story. The Confederation of British Industry has asked the Ministry of Technology to support efforts towards international alignment of standards and the acceptance of internationally agreed standards and to encourage BSI to accelerate the production of British Standards likely to be acceptable in overseas markets in advance of international agreement.'

Standards were indeed being deliberately geared to help the exporter. For example, a 1965 standard for fusion-welded pressure-vessels allowed higher design stresses than those in the long-standing specification for the British market – of considerable benefit to firms supplying export markets. New editions of standards for shell boilers and water-tube steam-generating plant incorporated ISO Recommendations and assured UK suppliers working to them of ready acceptance by insurance and inspecting authorities abroad. When in 1969 the British Standard for limits and fits was revised, it was decided to adopt in the new standard, BS 4500, a completely new presentation as a fully metric system in line with the corresponding ISO Recommendation.

In 1966 came the first moves which were to lead to a new bilateral agreement with Russia aimed at facilitating the exchange of information and flow of trade in precision instruments. Mr Binney led a team of British metrologists to Moscow and two years later Dr V V Boitsov, the Russian standards chief, brought a party to London on a return visit, the opportunity being taken to set up a joint UK-USSR working group on standards and metrology.

The IEC returned to London for its big annual session of grouped meetings in September 1968 – the first time for thirteen years. A thousand delegates from 29 countries were present and nearly 130 draft recommendations were completed during the meetings. Sir Stanley Brown, president of the Institution of Electrical Engineers, drew particular attention at the opening ceremony to IEC's work in the electronics field and the increasingly complicated problems of linking units and sub-units of equipment, sometimes supplied by different countries, in elaborate control schemes.

The change to metric

If international standardization was to overshadow all BSI's work in the 1960s, the change to the metric system was to be the most important single topic of consideration and activity.

In 1960, nine years after the report of the Hodgson Committee, the British Association for the Advancement of Science and the Association of British Chambers of Commerce jointly published a report entitled *Decimal coinage and the metric system – should Britain change?*. This report aroused widespread interest. As in the previous report, it was assumed that Britain could not make the change in isolation: in particular, without a parallel move by the United States, which at that time appeared most unlikely. But when the question was subsequently studied by the BSI Export Panel, set up under the aegis of the Executive Committee, its extensive soundings of opinion in industry made it clear that changes in Britain's pattern of trade no longer made it imperative for a decision by Britain to await a similar one by the USA.

Realization of this altered the entire case for metrication. From the point of view of current trends in international standardization, the argument for metrication was already convincing and from the standpoint of British exporters, the case for metrication had now also become much stronger. A clear consensus of opinion in industry favouring change, expressed through both BSI and the Federation of British Industries (now the CBI), led to the Government decision of May 1965 that industry on a voluntary basis should complete the change within a ten-year period, and that responsibility for planning and co-ordinating this operation should be placed on BSI, a reflection of the key role of standards in industrial production.

The implications of this decision had three major consequences for BSI. Firstly, the work of BSI achieved in this period a national significance which had previously been unparalleled except in wartime. The dependence of every sector of the economy on British Standards was brought home directly to all those involved in industry and commerce, whether as designers, production engineers, or purchasers, and many aspects of BSI's work in wider fields, such as medicine, education, and packaging, acquired a new significance. This development greatly enhanced BSI's position as a national body serving not only industry but every section of the community.

Secondly and in a similar way, the coincidence of metrication with the spectacular increase of international standards activities in the 1960s enabled the United Kingdom to play an even stronger part in the achievement of key international agreements during these years. These, in turn, made new initiatives possible in many fields where differing systems of measurement had previously constituted a formidable barrier.

Thirdly, the need to undertake the rapid revision in metric terms of British Standards, added to an already heavy work programme, led to an unprecedented rate of growth in the activities of the Institution. Reorganization and change became essential and, as so often happens in a crisis situation, necessity proved the mother of invention. Many of the new methods and procedures adopted in the second half of the decade can be closely related to these new exigencies.

BSI reappraises its role: The Feilden and Bowlby Reports

The early and middle 'sixties were to be a time for further reappraisal of working methods. With the Institution's resources at full stretch, there were serious worries about delays in progressing work and about finance. Some of these worries were pinpointed by the report in 1963 of the Government-appointed Feilden Committee on Engineering Design. The Committee's chief recommendation, so far as BSI was concerned, was that action should be taken to ensure that British Standards always encourage and never inhibit good design practice. Despite the general belief that British Standards represented the best practice and incorporated the most up-to-date knowledge, the Committee held that some standards did not meet these requirements. 'We would like to see BSI giving a more positive lead in the introduction of new standards incorporating the most up-to-date practice and in extending the field covered by British Standards.' Significantly, the Committee considered BSI 'insufficiently staffed for the immense task imposed on it by industrial society'.

In order to examine such criticisms and undertake a fundamental review of its role, the Institution set up in 1964 a panel under the chairmanship of Sir Anthony Bowlby, a leading Midlands industrialist and chairman of the Engineering Divisional Council.

In their report early the following year, the panel voiced concern as to whether the content of standards was always based on the right advice, particularly in those areas of industry where, thanks to research, rapid advances in design were taking place. They were anxious, too, about the time taken to prepare standards, pointing out the danger to British exports of any lagging behind in technical standards.

The Bowlby panel's broad conclusion was that the general basis of BSI's operation was sound but that a greater sense of urgency and wider recognition of the importance of its work were needed; and that the results of research must be more quickly reflected in standards. It urged industry to provide representatives on BSI committees who were expert from the commercial as well as the technical point of view: particular attention should also be given to securing advice from research and academic bodies.

The panel recommended more careful attention to priorities and the setting of target dates for completion of projects. It had strong views on how to speed up the preparation of standards and ensure the inclusion in them of design criteria. First, it endorsed for wider adoption a procedure which had already been tried experimentally, whereby technical committees considered draft standards only after they had been circulated for general comment by industry. Second, it held that the latest industrial techniques, if not included in the requirements of a particular standard, should at least be published in an appendix as a pointer to the future.

The Bowlby recommendations spelled out clearly that industry was looking to BSI to work more rapidly and decisively. What BSI now required was the wherewithal to forge ahead. A change in government at this time and the emergence of the newly-constituted Ministry of Technology, which became BSI's sponsoring department, provided the necessary new thinking on finance.

The role that BSI was already playing in assisting exports – international work occupying half the time of BSI staff – and the far greater role envisaged for BSI in the change to metric led the government to increase substantially its grant-in-aid. This, with rising revenue from subscribing members, enabled more well-qualified staff to be taken on. A central planning group was established to co-ordinate and programme the future workload that metrication would involve. From 1964 to 1969, in an unprecedented period of growth, the Institution's technical staff was, in fact, increased by 50 per cent. An even faster expansion rate was achieved by the Hemel Hempstead Centre, which in 1964 had celebrated its fifth birthday. Its staff rose in this time from 15 to 70, with five extensions to the original building (see Chapter 13)

An important development in 1966 was the establishment of the Technical Help for Exporters service, administered from the Centre and having as its objects the identification and elucidation of oversea technical requirements, arranging for testing or certification to overseas specifications, and the preparation of digests of information on particular groups or products.

To help accommodate BSI's expanding technical staff, the commercial departments (sales, accounts, subscriptions and membership records) had in 1966 been moved away from British Standards House to a new building near King's Cross which was named Newton House. From here BSI now distributes nearly two million copies of standards and codes annually. Opening the new building, Mr Wedgwood Benn, then Minister of Technology, noted that the government grant to BSI had been raised to match industrial subscriptions on a £ for £ basis and was being supplemented by a special grant for financing work on metric standards – over £400 000 of public money would be contributed to help BSI carry out this task. ' The success of BSI,' he said, ' will largely condition the success which this country can hope to achieve in technological advance. The most useful form of international technological collaboration is also the simplest – the acceptance of common standards.'

6

Standards to boost productivity : the standards engineer in industry

The preparation and publication of standards is one thing – that full use is made of them is quite another. Back in 1918 the British Engineering Standards Association included among its major aims the *promotion* of standards and this objective was underlined in the Royal Charter awarded in 1929: that the Association should promote the general adoption by industry of British Standards. Assuredly, once a standard is published there will almost certainly be some designers, manufacturers and buyers who will work to it or specify its requirements in their purchasing. But any systematic assessment of the extent to which standards are adopted by industry – and the economic results of their adoption – has, understandably, never been attempted; however in many areas, such as the cement and steel industries, the high coverage of production and sale to standards is not in doubt.

Since the end of the Second World War, efforts have increased, not only to encourage the use of the particular standards by those manufacturers immediately concerned, but also to encourage all firms to make better use of all the appropriate standards to eliminate excessive variety, boost productivity and generally streamline their procedures. Standards have in fact come to be recognized as an additional tool of good management. It is with this particular aspect of the promotion of standards, and the history of the development of BSI's Standards Associates Section, that this chapter is specially concerned.

This story really began in 1949 when a team was sent by the Anglo-American Council on Productivity to study the benefits reaped by United States companies from their war-time and post-war policy of simplification. The team, which included Mr G Weston, representing BSI, warmly recommended the extensive adoption of simplification by British firms and the major trade associations were asked to draw up proposals for elimination of excessive variety of types and sizes. ' Our dominant need today ' concluded the report ' is for a greater volume of goods at lower cost from the present productive resources '.

Standards engineers' first meeting

Larger companies in the United States, since well before the Second World War, had their ' standards ' departments aimed at deriving optimum benefits from mass-production techniques. Now in post-war Britain standards departments were also beginning to appear – or at least someone in the firm was designated to look after standards policy.

By 1955 standards engineers were sufficiently numerous and articulate for BSI and the Institution of Production Engineers jointly to organize a conference for them and this took place at British Standards House in May, with an attendance of over 100. From this first gathering, a committee of engineers concerned with standards matters was set up to advise BSI on the implementation and application of standards and to arrange future conferences.

These conferences, to become an annual feature of the standards scene, were notable from the start for the many and varied ideas thrown up. At this first meeting for example came the suggestion for ' summary sheets ' on the Continental model, containing extracts from standards, and this idea was to be put into effect within the next year.

Interest in using the techniques of simplification and standardization was at this time strong throughout Europe and indeed seminars for managements were held in the 1950s in various European countries under the auspices of the European Productivity Agency. A moving spirit in all this was Professor Harold Martin, on ' loan ' to the EPA from the Rensselaer Polytechnic Institute, New York, who was also the chief speaker at the second standards engineers' conference at British Standards House in 1956. He demonstrated how simplification and standardization had been applied in the United States and stated his clear view that the standards officer's function ' came well within the sphere of top management – in bringing control to bear on manufacturing processes, eliminating wasteful work and raising productivity, thus making possible higher earnings for workers and investors ' – music to the ears of this particular audience.

Two publications on the subject of simplification were issued at this time: a classic paper on *Variety reduction* by Professor Martin and *Some results of variety reduction*, consisting of case-studies, issued by the British Productivity Council.

At the 1956 standards engineers' conference the need to provide some form of training for standards practitioners was underlined and at the end of the year two courses were held, one at the Regent Street Polytechnic in London, the other at Manchester College of Technology – this being primarily for technical college staff, to provide material on standardization for introduction into their own courses.

The feeling that greater knowledge of standardization was needed was given further expression when, in April 1958, BSI and BPC set up a committee to encourage ' wider appreciation throughout industry as well as by education authorities, of the advantages to be derived from the three S's and particularly from variety reduction '. Its chairman was Sir Stanley Rawson, a leading industrialist and the chairman of BSI's Engineering Divisional Council.

The campaign against excessive diversity develops

The BPC launched in 1959 a year-long drive aimed at reducing unnecessary variety in manufacture, the opening shot being a conference held at the London headquarters of the Federation of British Industries (forerunner of the CBI). Mr Binney, in the opening address, reminded the audience that one of BSI's objectives was ' to eliminate the national waste of time and material involved in the production of an unnecessary variety of patterns and sizes of articles for one and the same purpose '. He referred to the growing awareness of the need for action in this field as evidenced by the holding of conferences, the creation of standards departments in firms and the increased emphasis placed on standards matters in technical education. ' What we are trying to achieve in all this is the rule of good order. Our object is to increase trade, to improve prosperity and to make more goods available.'

In various issues of *BSI News* attention was frequently drawn to the opportunities for reduction of unnecessary diversity following the issue of a new standard. Writing in *BSI News,* the sales manager for a company selling electric motors showed how the dice were increasingly loaded against ' specials ' with a basic cost on large production per motor of £11 compared with £44 for a one-off job.

How to organize a company standards programme

The third standards engineers' conference held in May 1957 attracted an audience of nearly 200 and a larger meeting place was found at Church House, Westminster. One of the most eagerly attended sessions consisted of presentations by two practitioners of how their standards departments worked. Obviously this was something about which industry badly wanted more information. Later in the year *BSI News* helped to fill this gap with a series of articles on how standards departments worked in a number of widely representative British companies – the main purpose being ' to provide a central pool of information which can be shared by other concerns of whom many are now taking their first tentative steps towards the adoption of a sensibly comprehensive standards programme '. Among the organizations who opened their doors to BSI for this purpose were Marconi, Rolls-Royce, British Thompson-Houston, Guest Keen and Nettlefolds, Fisons, the BBC, Kodak and many more. The then standards chief at AC-Delco was recorded as saying in May 1959 that it was not standardization for its own sake they wished to achieve but merely the elimination of uneconomic and unnecessary variety. ' There is no justification in, for example, designing 31 different gearbox take-off drives for speedometer flexible shafts, each one of which is eminently satisfactory in itself, when the end result could have been achieved with not more than four designs . . . we

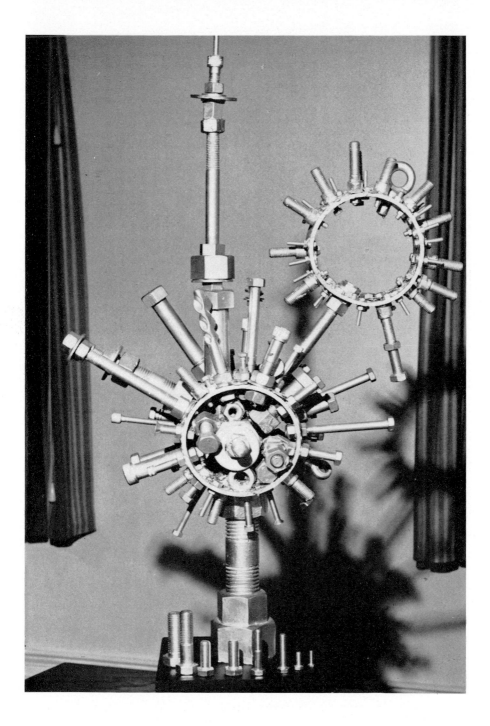

This 'pop art sculpture' mounted by Mather & Platt Ltd
in a 1969 metrication exhibition, illustrates how nine metric
fasteners (shown at the bottom) replace more than 60 types
previously used by the firm

[*To face p.* 38

Students at the first standards course held at Needham Hall,
Manchester, in July 1967, and organized by the Standards
Associates Section in cooperation with the Institute of Science
and Technology, University of Manchester (see page 40

cannot afford this superfluous luxury, neither can the motor industry or the nation.'

Similar ideas were expressed by Lord Halsbury, industrialist and scientist, when he opened the 1959 standards engineers' conference at the Connaught Rooms, in London.

Lord Halsbury urged management to encourage standardization. It was not just a question of 'telling George to keep an eye on standards matters' but of appointing an enthusiast and arming him with the necessary authority and facilities. It was typical, he said, for 30 per cent of a company's products to be non-standard and for these goods to yield a very trifling part of turn-over. It was known that these items were costly to produce but because 'it seems indecent to charge the customer all that' the firm sold these products at a discount. 'What they should do is to double the mark-up on such non-standard goods so as to discourage buyers' he argued.

In 1959 further help was provided by the publication of a booklet *The operation of a company standards department*, sponsored by a joint committee of the BSI and Institution of Production Engineers.

An association for standards engineers

Following suggestions at the 1959 and 1960 standards engineers' con-ferences for an association to provide opportunities for discussion nationally and regionally among those in industry concerned with standards, BSI set up in July 1960 its 'Register of Standards Associates'. A programme committee was formed to plan regional and other activities and the general management of the Register was put in the hands of the joint IPE/BSI Committee. By 1961 industrial membership was about 180; it has since increased steadily to over 500. The IPE/BSI Committee was reconstituted in 1962 as the 'Advisory Committee on the Use of Standards' with an extended membership. The Register became the 'Standards Associates Section' in 1963–4, under a more formal constitution and the various regional groups elected members to a programme committee given the job of planning the annual standards conference and administering the regions. In 1966 this committee became the Management Committee of the Section with the status of a BSI advisory committee and the BSI/IPE Advisory Committee on the Use of Standards was disbanded.

By 1964 the Standards Associates Section was in contact with similar bodies in other countries – France, Germany, the Netherlands, and these with the UK formed the nucleus of an informal group meeting from time to time to discuss common problems. Representatives from these and other countries often attend the annual standards conference now organized by the section, and UK representatives are invited to similar meetings on the Continent. (In the original discussion on the formation of the Standards Associates Section, the Standards Engineers Society of America which runs branches in other countries, including the UK and Canada, put forward the suggestion that UK standards engineers should be attached to it, but independence of the USA was preferred.)

The elected chairman of the section since 1962 has been Mr F E Butcher then group standards manager of Joseph Lucas Ltd. Mr Butcher received the OBE for his services to standardization in 1969 and has spoken on standardization as an instrument for management in USA, Russia, and other countries. The movement for standards engineers from all countries to discuss their common problems and exchange information has received impetus from the increasing importance of international standardization, and the Common Market negotiations.

From 1964 the Standards Associates Section worked towards the organization of formal training courses for standards engineers and prepared a syllabus for the course content which has formed a basis for the five courses so far held, with additions according to special university studies relevant to standardization or the application of new techniques of use to the standards engineer. The first course was held in collaboration with Manchester University Institute of Science and Technology in 1967, the second with the University of Strathclyde. Three further courses have been held in Manchester and in Bath. These courses have covered not only the broad field of national, international and company standards but also related subjects such as classification and coding, value analysis, group technology, and the computer as a tool for the standards engineer. As well as instruction to the participants they provide a very useful two-way exchange between industry and academics. All the courses have been filled to capacity.

Among the activities of the Standards Associates Section a particularly useful one has been the dissemination of information on metrication, and discussion of the problems as they affect different industries. Three standards conferences, those in 1966, 1969 and 1970, had metrication as their main theme. The 1966 conference was accompanied by a sizeable exhibition of metric products and a standards associate organized the Stocksbridge metric exhibition (see Chapter 7).

The original main function of the IPE/BSI Committee and now of the Management Committee, was to draw BSI's attention to the problems and needs of those in industry directly concerned with the application of standards. It brought in a homogeneous body of users whose point of view has often illuminated new aspects of standards work: references to compliance with British Standards in manufacturers' literature (now much more widespread); need for classification of lubrication oils (a project later carried out); requirements for bright steel, grades for sintered carbides, and other detailed technical points; sizes and lay-out of British Standards; need for better systems of retrieval of information contained in standards. In 1967 the Management Committee set up a Standards Application Panel responsible to it which took over the role of the former Advisory Committee on the

Use of Standards. The Panel's task is to investigate in depth so far as possible complaints and problems from individual associates in the regions on the operation of British Standards referred to it by the Management Committee and to prepare the submission to BSI on its behalf.

A major step forward in the integration of standards associates in BSI work was an arrangement agreed in 1969 for the Associates Section to nominate, when it considers it useful, an associate to a BSI technical committee to put the user's point of view.

In 1969 the Management Committee had considered the question of the professional status of the standards engineer and the decision was taken not to pursue the concept of a professional qualification in standardization bearing in mind that other professional qualifications are open to the Standards Engineer. Nevertheless a step forward in improving his position was taken when the Department of Employment asked the section to provide a definition of a standards engineer's functions for their occupational classification, under the heading of 'Professional and Related Occupations Supporting Management and Administration'. The definition provided underlined the opportunities for standards engineers to take part in management decisions and in reducing company costs. The important role of the standards engineer in industry has thus been given official recognition.

7

The momentous decision to change to the metric system

For many decades sporadic attempts had been made in Britain to encourage adoption of metric weights and measures. But it was not until after the Second World War that, face to face with the austere realities of postwar economic life, the country really began to take the idea seriously.

In 1951, the Hodgson Committee, appointed by the Board of Trade, firmly reported that a change to metric was desirable but that it should be made in concert with other inch-using countries with whom the UK traded.

Britain's position as an exporter having to deliver the goods to both inch and metric countries was the subject of lively debate at the fourth standards conference in 1958 when Mr H G Conway, a leading industrialist, suggested there was ' an atmosphere of inevitability ' about the adoption of the metric system. Conway had just been appointed to a study group set up by the British Association for the Advancement of Science to ' report on the practicability, implications, consequences both international and domestic, and the cost of a change-over to the metric system or the decimalization of weights, measures and coinage of the UK '. The British Association team, joined by representatives of the Association of British Chambers of Commerce, reported in June 1960, after taking the views of some 2000 organizations. Although the report found a strong case for the adoption of decimal coinage, on the wider issue of a change to metric weights and measures the Committees were unable to recommend any immediate compulsory change-over. They did, however, advocate the encouragement of decimal thinking to enable the UK to keep abreast of world trends.

On international standardization, the report's chief recommendation was: ' A common standard is ideal but when this is unattainable everything possible should be done to secure international acceptance of standards in both metric and inch/lb systems. There may be a case for BSI publishing further metric standards to facilitate manufacture in metric dimensions for export to metric countries .'

This highlighted the crux of the matter: the increasing problem of having to manufacture to two different systems of measurement. Faced with the problem in standards work, BSI decided that it should take soundings of its own. In 1961 the Export Panel, set up by the Executive Committee, initiated a special study. The Panel was fortunate in having as a member Mr C A J Martin, who had been chairman of the Association of British Chambers of Commerce Committee, and who was able to bring to bear his wide experience in the field of electrical engineering. A year later the main issues were clearly set out in a statement 'Change to the metric system?'. The statement, with a tentative 20-year programme for the change, was put before all the BSI industry standards committees for debate, industry by industry.

The temper of industry may be judged from the contribution to the 1963 standards conference by Mr C R Wheeler, the industrialist-chairman of the BSI Export Panel: 'At the present time the metric system is in use in all major industrial countries except the United States, Great Britain and part of the Commonwealth. About two-thirds of the world's population live in countries committed to the metric system. More and more of the emerging nations will wish to establish their own national standards and these are likely to be based on the metric system. How long can we as exporters afford to hold out against these odds?' he asked.

Overall industrial reaction to the BSI inquiry was published as *British industry and the metric system* in October 1963. A majority in industry recognized a change to the metric system as inevitable and considered that it should be made without delay, independent of any parallel action which might be taken by the United States or the Commonwealth. The report concluded that there was an 'almost unanimous desire for decision' and that indecision was acting as a curb on industrial progress. This recognition that Britain could, and, if necessary, should, act unilaterally marked a turning point in the century-old metric debate. From now on, events were to move swiftly.

In 1964 the Federation of British Industries (now the CBI) made further inquiry among its own extensive membership. The FBI conclusions were expressed in a letter sent by the FBI president to the government in February 1965: 'A majority in British industry now favours the adoption of the metric system as the primary system of mensuration for British industry, as soon as it can be brought about by general agreement'. The letter concluded that the time was 'appropriate for a general statement of policy on the part of the government, expressing support for the principle and giving some indication of the timing envisaged'.

The nation had not long to wait. In May 1965 came the historic statement in the House of Commons which finally set the country on the path of metres, litres and kilograms. Mr Douglas Jay, then President of the Board of Trade, said: 'The government are impressed with the case which has been put to them by the representatives of industry for the wider use of the metric system. Countries using that system take more than one half of our exports; and the total proportion of world trade conducted in metric units will no doubt continue to increase. Against that background the government consider it desirable that British industries on a broadening front should adopt metric units, sector by sector, until that system can become in time the primary system of weights and measures for the country as a whole.

'One necessary condition will be the provision of metric standards. The government have therefore asked the British Standards Institution, and the Institution has agreed, to pay special attention to this work and press on with it as speedily as possible.'

Focal point for the changeover

Mr Jay expressed the government's hope that within ten years the greater part of industry would have completed the change. In a letter to Lord Kilmuir, President of BSI, he wrote: ' . . . the international standardization work in which the Institution is playing such a prominent part can make a very substantial contribution to the expansion of international trade and not least to the development of our own exports. The decisions which we have now announced on the use of the metric system should be an important factor in the future conduct of this activity.' He went on ' . . . we attach the greatest importance to this operation; and we are very conscious of the central role which the Institution has to play in it '.

BSI's role was indeed central, since the rate at which standards could be revised in metric was the key factor in determining the rate at which industry could make the change. BSI thus became the pace-setter for metrication in industry, and the focal point for the co-ordination of all planning activities in virtually every sector between 1965 and 1969.

The first priority was to decide which of the 4000 standards in print were vital in metric to enable industry, particularly in the design field, to start making the changeover without delay. Standards covering the basic commodities used in production throughout industry, such as semi-finished materials of all kinds (rod, bar, wire, sheet, strip, etc), small components and measuring, cutting and shaping tools, formed the bulk of nearly 1400 standards selected for priority revision. Other standards such as glossaries, methods of test, chemical composition, were less urgently required, and could generally be metricated when they fell due for revision in the normal way.

But it was not enough just merely to develop a crash programme for the provision of metric standards. Unless industry was ready to use metric standards as they became available, there was a grave danger of widespread confusion and dislocation. The government's statement had amounted to little more than an expression of intent, together with a promise of general encouragement through support for BSI, and the use of public purchasing policy ' when practicable '. There remained a credibility gap to bridge before industry would commit itself to the necessary investment: many were sceptical that such a complex exercise would ever get off the ground. The real achievement of BSI during the next five years lay in securing the overall co-operation of British industry in working to an agreed timetable for the change.

The decision for British industry to adopt metric units was timely in

Mr A G Norman, the then President of the CBI, speaking at
the publication conference for the BSI programme for metrication in engineering, July 1968. Also in the picture (from
left): Mr E W Greensmith, Chairman of the metrication
policy committee responsible for the programme, Mr
Anthony Wedgwood Benn, Minister of Technology, Mr G H
Beeby, Chairman of the General Council, Mr H A R Binney,
Director General, and Col J S Vickers, Head of the Planning
Group. The BSI metric 'key' symbol is featured in the
background (see page 45)

[To face p. 44

The adoption
of the
metric system
in engineering:
Basic
programme
and guide

PD 6424
July 1968
15 shillings

Published by the British Standards Institution

The programme for metrication in engineering, published by BSI in 1968, formed the basis for more detailed planning throughout industry. Programmes adopted in other countries changing to the metric system followed it closely

THE ADOPTION OF THE METRIC SYSTEM IN

ENGINEERING: BASIC PROGRAMME

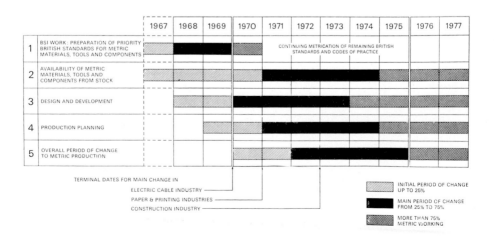

		1967	1968	1969	1970	1971	1972	1973	1974	1975	1976	1977
1	BSI WORK: PREPARATION OF PRIORITY BRITISH STANDARDS FOR METRIC MATERIALS, TOOLS AND COMPONENTS					CONTINUING METRICATION OF REMAINING BRITISH STANDARDS AND CODES OF PRACTICE						
2	AVAILABILITY OF METRIC MATERIALS, TOOLS AND COMPONENTS FROM STOCK											
3	DESIGN AND DEVELOPMENT											
4	PRODUCTION PLANNING											
5	OVERALL PERIOD OF CHANGE TO METRIC PRODUCTION											

TERMINAL DATES FOR MAIN CHANGE IN
ELECTRIC CABLE INDUSTRY ———————
PAPER & PRINTING INDUSTRIES ———————
CONSTRUCTION INDUSTRY ———————

INITIAL PERIOD OF CHANGE UP TO 25%

MAIN PERIOD OF CHANGE FROM 25% TO 75%

MORE THAN 75% METRIC WORKING

another sense for it followed international agreement on new rules for the use of the metric system. This rationalized and simpler version, known as the International System of units (SI), has many advantages. By deciding to move straight to it, British industry was placing itself at an advantage relative even to that of traditional metric using countries, virtually all of which were now committed to the adoption of the SI. Appropriately, a booklet on the use of SI units published by BSI in 1966 became one of the first major guidance documents in the changeover. Over 100 000 copies were sold in the following five years.

Planning the change

The construction industry was one of the first to prepare detailed plans for metrication – plans which were able to take account of the introduction of dimensional co-ordination into standards work. In April 1966 BSI circulated a questionnaire to the industry; the results of this provided the basis for a national programme published by BSI in February of the following year. With 1972 set as the target date for substantial completion, the construction industry was clearly setting the pace.

In engineering and other sectors, the unique opportunity provided by metrication for re-thinking existing standards in the light of industrial needs, reducing unnecessary variety wherever possible, was not lost. The idea of metrication as an unlooked-for opportunity for improvement gained rapid acceptance throughout industry, and many companies were soon reporting major savings, partly springing from variety reduction techniques but just as much from the radical re-thinking which the programme encouraged.

In April 1966, a central planning group was formed under the direction of Col J S Vickers to co-ordinate BSI's metric activities. Over the next 18 months the group's chief task was the preparation of an outline changeover timetable for the engineering industries on which virtually all sectors depended for materials and equipment.

Extensive consultations with trade associations and individual companies preceded publication of the final programme for engineering in July 1968. This provided the framework for almost all subsequent planning activity, and in January 1969 BSI published related programmes for the marine and electrical industries. Throughout the year 1969, more detailed programmes were worked out through the efforts of some 80 trade associations covering nearly every sector of industry.

Parallel to this work, BSI launched major publicity and educational campaigns directed at industry. By 1970, BSI had provided speakers for more than 400 outside lectures covering about 40 000 people, over and above those given by the construction industry's speakers panel. A leading role was played, too, by the Standards Associates in organizing metrication conferences and seminars. One special exhibition organized in 1966 by the

Associates in Stocksbridge, a small Yorkshire town at the heart of the steel industry, aroused widespread national interest.

To publicize the changeover, BSI introduced a special 'metric key' symbol, which quickly became established as the national symbol for identifying metric products and literature. The key symbol was used on all BSI metrication literature which, together with a mobile exhibition and other BSI metrication aids, were in constant demand. The information departments of the Institution were swamped with inquiries from companies and other bodies who were now planning their conversion programmes.

Decision for the non-industrial sectors

As interest in metrication spread to other sectors, such as medicine, education and the consumer field, BSI was drawn into wider planning activities. Especially noteworthy was the part it played in aiding the Royal Society in its metric initiatives in the education field. Two major conferences were held by the Royal Society in 1968, bringing together representatives of the teaching profession, educational suppliers and publishers, and educational authorities. In the field of technical education, the Council of Technical Examining Bodies and the Construction Industry Training Board played a similar planning role.

Following its 1965 announcement, the government had set up a Standing Joint Committee on Metrication to consider the implications of metrication for the country as a whole. Both BSI and CBI stressed to the Committee the urgent need for parallel conversion programmes for the retail and other non-industrial sectors. When, in 1968, the Committee made its report, it recommended that the government should set the end of 1975 as the target date for the adoption of the metric system by the country as a whole and that a Metrication Board should be set up to stimulate and co-ordinate sector planning.

The Metrication Board was set up in May 1969. The new Board operated through eight steering committees, on each of which BSI was represented. Despite the difficulties inherent in assuming responsibility for co-ordinating programmes that were already well under way, it was not long before the Board had assessed what still needed to be done. However, a change of government after the 1970 general election produced uncertainty about official policy on metrication, and, with public attention focused on decimalization of the currency in February 1971, a lull in planning for the retail sector followed.

In the meantime, the case for metrication was considerably strengthened by developments in other parts of the world since 1965. First South Africa, then in close succession Ireland, New Zealand and Australia announced their intentions of changing, and established the necessary machinery for this. Britain's decision was undoubtedly a major factor in bringing these changes about, and in each case the programmes closely resembled those adopted in the UK.

A further development was the decision taken by Canada in 1970 to make the change, and by the middle of 1971 virtually every other country in the Commonwealth was committed to metrication. The United States remained as the sole major imperial using country and even here the tide was running strongly in favour of change. In August, the Secretary of Commerce recommended to Congress that the US should adopt a ten-year conversion programme. His report to Congress, referring to Britain's experience, stated ' their metrication effort serves as our pilot programme '.

The historic decision taken at the end of October 1971 that the United Kingdom should enter the European Common Market produced a final – and conclusive – argument in favour of metrication. This was recognized in the White Paper on metrication published in February 1972: ' the Government believe that the time has now come when they must act to ensure the orderly completion of the process '.

8

The origins of international standardization

Practically every industrial country in the world has its own standards body and firmly established national standards which invariably influence the acceptance of imported goods. In today's export conscious world there cannot be many in industry who do not see that standards must therefore, as far as possibly, be international in outlook. As the years have gone by, BSI, as the standards body of one of the most export-conscious of all countries, has played an ever-increasing part in the process of dismantling the standards barriers to trade.

The foundation of BSI preceded by nearly 20 years that of the other major national standards bodies. The United States did not have any comparable body until 1918, when the American Standards Engineering Committee (now the American National Standards Institute — ANSI) was formed. In Germany, the Standards Committee of German Industry (now the Deutscher Normenausschuss — DNA) was established in 1917, and in France the creation of the Permanent Standardization Committee in 1918 preceded the foundation of the Association Française de Normalisation (AFNOR) in 1926. In the years before the First World War, the British position in trade and finance tended to be a dominant one. Then, and for many years later, little thought can have been given by our predecessors to international standards: British Standards themselves could be regarded as hallmark enough.

Thus, it is hardly surprising that with one or two notable exceptions the early efforts by the British Engineering Standards Committee in this direction were concentrated on promoting the use of British Standards by our major trading partners, especially within the British Empire and in the encouragement of local standards committees modelled on the British pattern. Only where this was not possible were wider efforts made to secure agreement, as in the early efforts made from 1910 to secure agreement with the USA on screw threads. The most important exception to this was in the case of the electrical industry, a comparative newcomer to the industrial scene. That industry provided the first real step in the promotion of international standardization as we now know it, in the founding between 1904 and 1906 of the International Electrotechnical Commission. This stemmed primarily from the initiative by Professor Elihu Thomson for the USA and Colonel Crompton for the UK at a meeting in St Louis in 1904. Colonel Crompton was accompanied at that meeting by Mr le Maistre, then a young electrical engineer serving with the British Engineering Standards Committee. This was followed by a gathering in London in June 1906 which provided the rules for the 'International Electrotechnical Commission', when the great Lord Kelvin became the first president of the IEC, with Mr le Maistre as general secretary. Mr le Maistre (later the first director of BSI) served continuously in this post of IEC general secretary until his death in 1953.

A point to be noted here is that the IEC was formed in advance of the creation of national standards bodies round the world, except for the case of BSI. It is as a consequence of this that in many countries the responsibility for international work for electrical standards has been divided from the responsibility of work in other fields. When in the years after the First World War the other major national standards organizations were established, there were already separate national electrical technical committees to take care of the IEC work, whereas from the start the UK has had a single, centralized organization. The pattern of BSI in providing a single point for all of this country's standards work has, however, been followed in all those countries where standards bodies have been set up following the Second World War, as well as in a few of earlier origin, and in all those others in the old Commonwealth which have largely modelled themselves on BSI.

In April 1926, the 18 countries which by that time had set up standards bodies met together to consider the extension of international collaboration to other fields. As a result, the International Federation of the National Standardizing Associations, known as the ISA, was formed, with most of its strength from Europe. But when in 1927 the British delegation under Sir Richard Glazebrook reported on their discussions, the opportunities offered by association with the ISA were not taken up by British industry, and this country did not join effectively in international standards work until after the Second World War, except in the important sphere of the electrical industry.

The ISA had a not particularly eventful life of 13 years. It ceased to exist when war broke out in 1939. In 1944, following Anglo-American discussion on the coordination of allied standards for war purposes and post-war needs, a standards coordinating committee was set up under the United Nations banner – with Charles le Maistre as the natural choice for its acting secretary. The Committee met in New York in November 1945 and arranged a full-scale conference in London in October 1946 at which representatives of 25 standards bodies agreed to set up the International Organization for Standardization – ISO. The IEC became its electrical division, though with full technical and financial autonomy.

Standards committees overseas

From the start, there had been a strong awareness of the need to promote the adoption of British Standards overseas – some of the earliest British Standards were for Indian Railways. In 1910, Sir John Wolfe Barry had proposed the setting up of standards committees abroad to promote the interests of British engineering and these local committees were to become the forerunners of many national standards organizations, generally working to the pattern established in the UK.

Among the former Dominions, South Africa was the first in 1908 to have its standards committee, with the South African Bureau of Standards founded in 1934. In 1919 a standards body was formed in Canada, which in 1944 became the Canadian Standards Association. Also, in 1919, a standards committee was set up in India, forerunner of the Indian Standards Institution, founded in 1946. Pakistan formed its own standards institution in 1950. In New Zealand a local committee of the British Engineering Standards Association was formed in 1920, and became the Standards Association of New Zealand in 1932. The first step in Australia was the foundation in 1922 of the Commonwealth Engineering Standards Association to become, five years later, the Standards Association of Australia. Standards bodies were also formed by Ireland and Israel (then Palestine) before they left the Commonwealth. More recently, BSI has been closely concerned with the establishment of standards bodies in Central Africa, Malaysia, Ceylon, Trinidad and Jamaica.

From 1910, with help from the Foreign Office and the Board of Trade, efforts were made to make British Standards known in foreign countries also. In 1918, a committee of British engineers was set up in Shanghai, and in South America, where there had long been considerable British trading interests, a special British Standards committee was established in Argentina in 1936 and continues to this day as a liaison organization with the national standards body and as a sales and information centre for British Standards.

The Commonwealth Club

Naturally enough, it is in the Commonwealth countries that British Standards have been most closely followed. In Australia and New Zealand, British Standards are still often adopted as they stand, or are modified to suit local circumstances.

The pattern was set at the Imperial Economic Conference in London in 1930 when it was recommended that the standards bodies of the Commonwealth countries should maintain regular consultation with a view to establishing uniform standards. It was not until after the end of the Second World War that the first in an important series of independent Commonwealth Standards Conferences took place. It was held in London in 1946 with eight countries participating: Australia, Canada, Eire, India, New Zealand, Palestine, South Africa and the UK. The main items on the agenda were the development of certification marking and the influence of the Commonwealth standards bodies in the wider field of international standardization. Those taking part could note with some satisfaction that the democratic principles

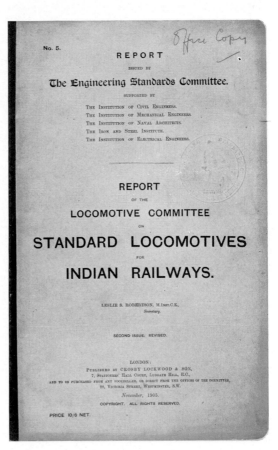

One of the first British Standards,
BS 5, was for locomotives for
Indian Railways (see page 49)

THE ENGINEERING STANDARDS COMMITTEE.

PASSENGER STANDARD ENGINE & TENDER
FOR
METRE GAUGE

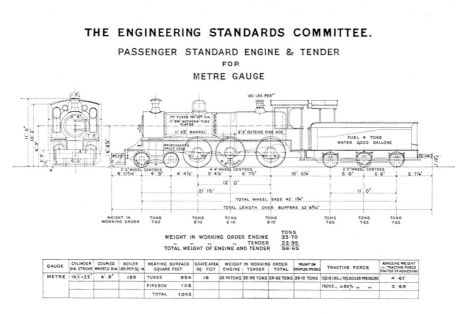

GAUGE	CYLINDER DIA. STROKE	COUPLED WHEELS DIA	BOILER LBS PER SQ. IN	HEATING SURFACE SQUARE FEET		GRATE AREA SQ. FEET	WEIGHT IN WORKING ORDER			WEIGHT ON COUPLED WHEELS	TRACTIVE FORCE	ADHESIVE WEIGHT ÷ TRACTIVE FORCE (FACTOR OF ADHESION)
							ENGINE	TENDER	TOTAL			
METRE	15½ × 22	4′.9″	180	TUBES	954	16	33·70 TONS	22·95 TONS	56·65 TONS	26·10 TONS	12518 LBS @ 75% BOILER PRESSURE	4·67
				FIREBOX	108						15022 ,, @ 90% ,, ,,	3·89
				TOTAL	1,062							

Scale $\frac{1}{8}$ Inch = 1 Foot.

[To face p. 50

Mr Binney in conversation with Mr P J Ailleret, President
of the IEC, September 1968

so far-sightedly established by the British pioneers at the beginning of the century had been adopted by all the Commonwealth successors – notably that standards should be arrived at by general consent, that they should fill a recognized need and that they should preserve community of interest between producer and consumer.

A second Commonwealth conference in 1951, also in London, was held in conjunction with the BSI golden jubilee celebrations. Problems of compliance with varying safety standards and regulations in the different countries were specially noted and it was agreed that a study of domestic electrical appliances in this context should be undertaken.

In 1953, reflecting a desire for more detailed discussion on technical matters, as a way of minimizing differences between standards in the countries concerned, a conference on cables was held in London. When the next full-scale conference took place in Delhi four years later, a particular achievement was the acceptance of BS 2771 on electrical equipment of machine tools by all the other Commonwealth countries as a basis for their own standards. After a meeting in Ottawa in 1962 the discussion of standards for particular fields, in full technical sessions, was not continued because the rapidly advancing work in ISO and IEC had made it superfluous; but it was generally agreed that Commonwealth agreements on specific standards topics had exerted a beneficial influence on the wider international discussion then getting under way.

At the 1957 conference, delegates had an opportunity to discuss at first-hand the effects of India's change-over to the metric system. The 'club' agreed then to work as far as possible towards interchangeability between products made to inch and metric dimensions.

At subsequent conferences in Sydney in 1962 and London in 1965 the discussions veered more towards such matters as consumer goods standards, certification marking, information procedures and so forth. At the 1965 conference there was a much larger attendance than ever before with representatives from the newer Commonwealth countries as well – Central Africa (Rhodesia), Hong Kong, Jamaica, Malta, Mauritius and Nigeria. South Africa, although having left the Commonwealth, remains a participant in these consultations – so does Eire. It was a good opportunity to discuss ways in which the older-established standards organizations could help the newer ones. Other subjects discussed were Britain's change to metric, the need to speed up the preparation of standards and achieving a common lay-out for them, and the development of informative labelling.

America-Britain-Canada achieve unity

The screw-threads saga was still far from being concluded at the beginning of the Second World War, but the pressures on those concerned to find a solution were stronger. By 1943 it was recognized on both sides of the

Atlantic that serious delays in war production were resulting from lack of screw-thread agreement between the allies. But even then it was not until the war's conclusion that there came from a conference in Ottawa in 1945 the famous ' Declaration of Accord ' through which the Americans, British and Canadians agreed on unified screw-thread standards.

From this agreement stemmed much other ' ABC ' work on unification of basic engineering standards – limits and fits, drawing practice and metrology. In 1955 other Commonwealth countries were invited to send observers to an ABC meeting in London. Two years later, in Toronto, came another notable agreement, on drawing office practice, ' to eliminate any significant difference in principle so that drawings prepared in accordance with the American, British or Canadian standards – when amended – might be understood and used in the factories of each country '.

ABC work, as with standards collaboration among the Commonwealth countries, has become somewhat less essential as many of the subjects concerned have been taken under the wing of the world standards body – the ISO (Chapter 9). So it was that by 1960 the ABC countries were able to agree on arrangements for co-ordinating their efforts in the wider ISO sphere, although further ABC meetings have taken place to ensure a common approach in the world forum.

9

BSI and its world neighbours in the ISO and IEC

ISO from its birth to the present day

As was noted earlier (page 49) BSI had played a large part in the preparations for setting up the new International Organization for Standardization which was created in 1946. All concerned in trade, not just those in BSI but managers and technicians in many industries, encouraged by government, were now determined to support this promising new venture. There was to be none of the lethargy and distrust which caused the earlier ISA effort to fail.

The new ISO was given headquarters in Geneva and its first governing Council consisted of the standards chiefs of the United States, Russia and China, the UK and France as well as Australia, Belgium, Brazil, India, Norway and Switzerland. This was from its inception, continuing through various periods of the cold war, a truly international body. While statesmen have ranted, the technical committees of the ISO have regularly and frequently come to amicable agreement on a host of subjects destined to ease trade between country and country.

In 1949, some 50 technical committees—taking up in some cases where the ISA had left off—had already been established in the ISO covering pretty well all the basic and long-established activities and industries: screw threads, steel, machine tools, automobiles, boilers, textiles, as well as newer ones such as aircraft. BSI had taken secretariat responsibility for ten of them; BSI was also active from the start in special planning and editing committees. One such special committee, set up in 1953, was the Committee on Scientific Principles of Standardization which concerned itself with such matters as the reconciliation of inch and metric dimensions, implementation of the SI, and the economic effects of standardization.

By the mid-1950s the 'recommendations' which had come out of ISO committees were beginning to have their effect on trading practices and more industries were pressing to have their standards problems examined in an international forum. ISO now had 80 technical committees and the pace of work was accelerating so rapidly that the load on the secretariat in Geneva and delays in processing documents there were causing concern. BSI was among those pressing strongly for a speeding up of the work and a general strengthening of the secretariat. There followed a 50 per cent increase in subscription of the member-countries in 1957, and Mr Gordon Weston of BSI was given the task of reviewing the secretariat's whole method of working. Two years later, with growing interest in international standardization by the developing countries – ISO membership had risen to 40 – the Council was enlarged to 14, the extra places making room for representatives of the newer countries. Other ways of helping the emergent countries have been the setting up of a special ISO committee to tackle this problem, a form of 'correspondent' membership at a reduced rate of subscription, organization of seminars, and training help from the established standards bodies. In 1968 a co-ordinating bureau was set up consisting of the ISO and IEC secretariats together with UNESCO and UNIDO; this has responsibility for giving general advice on the promotion of standardization in developing countries and in training programmes.

The use of an ISO Recommendation, as it stands, has not been the practice; the procedure being for each of the standards bodies to take account of international recommendations in its own national standards – thus leaving scope for any necessary variation or refinements. There has, however, been increasing pressure to avoid conflict between national standards and the requirements of international recommendations. More recently complete adoption of international recommendations as world standards has become the recognized idea.

Growing interest in the implementation of ISO Recommendations was becoming specially apparent from 1960 onwards, when it was urged that all member-countries should indicate in their catalogues whether their standards were in line with the corresponding ISO Recommendation. In 1964 an ISO survey showed a high correlation between ISO Recommendations and national standards – particularly in the UK, France, Czechoslovakia, USSR and Germany.

About this time concern was again felt about the Geneva secretariat not being able to keep pace with the technical work and a committee under the chairmanship of Mr Binney was set up to review the secretariat's requirements and advise on recruitment of staff. Considerable expansion took place in 1966–67 with a view to speeding the output of ISO Recommendations. A single Executive Committee with Mr Binney as its chairman was set up in 1967 and ISO's technical committees were divided administratively into four groups of related subjects, with one member of ISO's staff attached to each to ensure greater co-ordination.

From an annual rate of a handful of publications issued each year, ISO's output has gradually crept up and is now in the region of 400 a year. Appointment of a public relations officer and the publication of a magazine is helping to promote the interests of ISO to a wider public. An ISO technical information centre was opened at Geneva in 1971.

Today ISO is the largest international system of industrial collaboration in the world, constituted of 55 member countries. BSI continues to play a leading role and hardly a meeting of the 146 technical committees takes place without UK-BSI representation, and BSI holds the secretariat of a quarter of them. At the most recent ISO general assembly – at Ankara in 1970 – the general secretary, Mr Olle Sturen, noted that they had published over 1400 recommendations, two-thirds of them having been issued in the previous three years: the successful outcome of the previous years of initial work and special streamlining efforts. An interesting development has been a recent decision that ISO should issue ' standards ' in place of ' recommendations ', with effect from the beginning of 1972. This is a considerable psychological victory for internationalism in standards work.

Into the consumer field

By the mid-1960s ISO had established liaison with the International Office of Consumer Unions and a statement was issued by ISO emphasizing the importance of a close link at national level between standards organizations and consumer bodies.

In 1966 an ISO advisory committee on consumer questions laid down that international and national standards organizations should give the fullest possible help to consumer bodies in supplying shoppers with clear and objective information about products through standards, informative labelling and comparative testing. By 1968 the committee was closely concerned with programmes for standard methods for measuring the performance of various types of consumer goods – linoleum, carpets, gas cookers, record players and lawn mowers, among others. At a meeting in London in 1969 the committee was told by Miss Eirlys Roberts on behalf of the International Organization of Consumer Unions that speed was vital and that projects helpful to consumers should not be held back because the final technical solution to a particular problem was not yet available. An International Steering Committee for Consumer Affairs, with representatives of ISO, IEC, and consumer associations, was set up in Geneva together with a high-level committee for certification matters.

In 1970 came a first meeting of a new ISO committee on sizing systems for clothes, which agreed that the sizing designations should relate to body, not garment, measurements.

The ISO, like so many of its constituent national standards bodies, has felt obliged to became involved with matters directly affecting the man-in-the-street. But, of course, it is the technical sphere that remains the ISO's main area of activity.

Some ISO achievements

The range of ISO work is so immense that it is impossible to do justice to it in a page or so. But here are a few examples of achievements which have come out of long and patient discussion in its technical committees – in particular, the achievements in which the UK has been specially concerned, and which have had, or are having, a significant effect on world trade.

Steel This committee, of which BSI holds the secretariat, first secured international agreement on methods of mechanical test of steel in its various forms – plates, bars, sheet and strip, wire and tubes, etc. Another important contribution has been the development of methods for the high-temperature testing of steel, an important consideration for those sections of industry concerned with production and export of boilers and other high-temperature and high-pressure apparatus operating at temperatures up to 1000 degrees C. The committee has gone on to issue recommendations for steel sections, qualities of structural steel, heat treated steels and other steel products. There are now some 60 ISO Recommendations in this field.

Nuclear energy This committee, to whose work Britain has made a substantial contribution, has been able to evolve international agreements before national practices have become too rigidly set. It is concerned with terminology, definitions and units; radiation protection; radioisotopes; and reactor safety.

Machine tools The aim of this committee has been to develop a test code for use as a standard of reference throughout the world. It has also been responsible for the preparation of a number of recommendations dealing with such items as lathe tool posts, self-holding tapers for tool shanks, tapers for tool shanks, lathe centres, tee slots, symbols on machine tool indicator plates, spindle noses and face plates.

Textiles World agreement has been achieved on numerous methods of test for fibres, yarns and clothes – many of which have an important bearing on commercial trading. For example, an agreed procedure for ascertaining the weight of fibres and yarns is used as a basis for fixing prices; reduction in the variety of cloth widths helps simplify mill production.

Hypodermic syringes and needles This is one of the ISO committees with a strong humanitarian motive. Functional interchangeability of fittings used to connect the needle to the syringe is a main objective – to allow for matching together of parts from different countries; other advantages of this work result from agreed methods of size designation and simplification of size ranges of needles. Similar work has been undertaken in regard to anaesthetic equipment.

Personal safety BSI took the lead in setting up this committee which, using sound British Standards as a basis, has developed international recommendations now widely agreed throughout the world for protective helmets and car seat belts.

Fire tests With the rapid development of new building materials and constructional methods it has become increasingly necessary to have adequate tests to apply to materials and structures to assess their behaviour in the event of fire. This committee, with a BSI secretariat, has worked towards international agreement on such tests.

Acoustics To help combat the problem of noise this committee has laid down recommended methods for measuring noise – from motor vehicles, aircraft in the vicinity of airports, industrial machinery. This committee also issued in 1955 one of the earliest of ISO Recommendations of world-wide

The ISO Technical Committee for Photography, ISO/TC 42,
in session at BSI's Conference Centre, Hampden House, 1972

[To face p. 56

Members of the IEC Council and Committee of Action at
the XXXIII IEC General Meeting held in London at BSI's
Conference Centre, Hampden House, in September 1968
(see page 31)

significance – that for 'musical pitch' – based on the British Standard.

Computers The BSI committees concerned with computers and data processing work very closely with the equivalent committees in ISO. This subject is one in which it is common for an initial agreement to be worked out in the international committee so that it may be adopted in the national standards of the member countries. Occasionally, however, one country or another prepares standards in advance of the international work and the standards which BSI has prepared in this category include those for digital input/output interface for data collection systems, safety of office machines and data processing equipment, and fire protection for electronic data processing installations.

Ball and roller bearings This, one of the first ISO committees to start work, has done much to achieve world-wide agreement on standardization of basic engineering components. By reducing the number of types and sizes manufactured, it has helped to concentrate production and bring down costs. International standardization of boundary dimensions and tolerances has been a major contribution to world-wide trade, and means that fully interchangeable bearings can be provided the world over for machinery of every kind.

Boilers Over the years countries had issued statutory regulations governing the design and construction of boiler plant, and since different countries held different views on the necessary requirements, international trade had become complicated. This ISO committee has worked with much success for world agreement on a single international boiler code, endeavouring to unify the various national codes and give a lead to those countries which have not yet drawn up codes of their own.

Freight containers Exporters and importers the world over appreciate the need to speed up delivery and cut the costs of freight handling. When this committee began work in the early 1960s there were no international standards for freight containers and merchandise was dispatched in containers of various sizes and capacities. This meant that a container carried by ship might be too large to carry on a road vehicle or train. Different countries had different transport regulations limiting the size and weight of the loads which might be carried. The committee has developed a range of sizes which are fully interchangeable between one type of transport and another, the world over.

International standardization of electrical equipment: the IEC

The work of the International Electrotechnical Commission, whose beginnings in 1904 were noted in Chapter 8, has continued steadily and fruitfully, interrupted only by the two world wars. A great body of technical agreement has been built up on the specification of components and products which go to form industrial electrical contracts. Organized by 'national committees' in each of the 41 member-countries of IEC, the Commission's recommendations were originally of the definitions-and-tests variety rather than detailed specifications; this approach has, in fact, changed considerably in recent years. The IEC catalogue now contains over 400 published recommendations.

After the interruption of the Second World War, the IEC Council met again in July 1946 in Paris and within two years a dozen of the technical committees – each of which is responsible for a major sphere of activity: radio-communication, switchgear, cables, etc. – had contrived to get the work moving again and arrange meetings. During the 1950s and 1960s the tempo of activity steadily grew.

Every year from 1948 a large-scale 'general meeting' has been held, a different country playing host on each occasion. At the 1954 general meeting the IEC celebrated its fiftieth anniversary when some 50 technical committees met in Philadelphia. The strong British delegation was led by the statesman Lord Waverley who was then president of BSI. Increased interest and participation in IEC work by the United States was greatly welcomed.

It was at this meeting that the IEC's Committee of Action expressed concern at the absence of international recommendations on general safety of electrical equipment and urged the preparation of IEC recommendations covering fundamental aspects of safety. In 1955 a special sub-committee was set up under the chairmanship of Mr Binney, Director of BSI and a member of the Committee of Action, to examine the problem, and at the same time technical committees were asked to expedite work on fundamental safety. An Advisory Committee on Safety Rules was appointed in 1956 to keep general safety problems under review. In 1959 it prepared directives for the technical committees, giving guidance on safety features and a priority list of equipment for which safety recommendations were needed. The subject was again to the fore in 1962 when the possibility of a general safety code was considered, but it was thought that this was too ambitious a project and that the only practical step was to get agreement on the requirements for particular classes of equipment. In 1963 the Advisory Committee defined classes of low-voltage equipment according to their means of protection against electric shock, and prepared a list of safety clauses from which technical committees could select points appropriate to their particular work.

In 1955 work on electronics was growing in importance and a special committee was asked to make recommendations, a permanent advisory committee on electronics being established in 1960. By 1961 the IEC had twelve technical committees and 15 sub-committees at work in the electronics and telecommunication field, with the UK and United States particularly involved. 1956 saw the start of IEC work on nuclear energy – measuring instruments for nuclear reactors, components for servo-mechanisms and measuring instruments for radio isotopes. International standardization of flameproof enclosures was another subject for strong

UK participation: IEC recommendations were published in 1955, followed by consideration of alternative methods such as oil immersion and intrinsic safety; in 1958 it was agreed to review the possibility of a single IEC publication classifying types of protective enclosures. In 1959 came IEC involvement in standards for automatic data processing.

The question arose in 1962 of the relative spheres of IEC and CEE (see below) in regard to domestic appliances, CEE at that time contemplating work on performance as well as safety of domestic appliances. No immediate decision was made, but in 1964 IEC set up its own committee for performance of domestic appliances – reflecting national and international consumer demand for agreement in this field as the basis of informative labelling arrangements.

Another major development in 1967 was the setting up of a committee on electrical control of processes – a quite new concept in IEC work with the committee empowered to consider complete electrical measurement and control systems, leaving the specification of individual pieces of equipment to other, specialist, committees.

Two years later came the start of work on fundamental and far-reaching international standards agreements. First a decision to compile an international set of wiring regulations for buildings – to facilitate trade between countries (which has not infrequently been hampered by differences between national regulations), and to provide guidance as to good practice for emergent countries; the international regulations promise to be similar to those of our IEE but will probably go further. The second was an IEC decision to make another attempt to standardize a plug and socket for international use – a difficult task but one which electrical engineers and users the world over would greatly appreciate being achieved.

There are many common problems and technical areas where the responsibilities of ISO and IEC overlap, more as electrical and electronic techniques enter into other equipment. Much of the liaison is *ad hoc*, between the ISO and IEC secretariats, sharing the same building in Geneva, and between technical committees. A general co-ordinating committee was set up in 1966 to assist these liaisons.

The CEE and its work to unify standards for domestic electrical equipment

An organization bringing together testing and approval authorities, and known as the Installationsfragen Kommission, had been set up in 1926 with a view to harmonizing European standards for domestic electrical equipment. It was reconstituted after the Second World War as the International Commission on Rules for the Approval of Electrical Equipment, known more familiarly as the CEE. This body to which 19 European countries belong – east as well as west – and to whose meetings other countries including the United States sends observers, deals in particular with safety aspects of standardization and is mainly concerned with equipment used in the home. In fact the long-term objective of the CEE is the very desirable one of ensuring that countries having official approval schemes for electrical appliances should base their approval on identical specifications. A licensing scheme is operated which provides for the reciprocal acceptance of test certificates by approval bodies in each of the member-countries (Chapter 10).

One aspect of CEE work which has been having a quite profound effect in millions of European households has been the agreement over identification colours of flexible cords. Discussion began in 1957, with the CEE proposing green/yellow for the earthing core in place of the various single colours in use in different countries The British Standard colour had always been green; in some countries it was red – an obvious danger with imported appliances since red had always been our colour for ' live '.

All CEE members agreed in 1960 to accept the green/yellow compromise either as the sole national standard or as an alternative to the existing identification colour.

From 1964 to 1966, on Britain's initiative, further thought was given to the alignment of the other core colours, for neutral and live. Britain fought hard to retain red for live but this was quite unacceptable to Germany and other countries which had long used red for earth. At one stage Germany volunteered to accept black for neutral (the UK practice) as a bargaining point, but again this was not acceptable to other countries. Eventually after much argument all 19 countries agreed in 1967 on light blue for neutral and brown for live in the case of three-core flexibles – no mean victory for the spirit of compromise which must of necessity permeate all the thinking in international standards discussion. The agreement was far-reaching in its consequences, demanding changes in legislation in some countries, and much education and publicity in all of them.

International Special Committee on Radio Interference

This special committee, known as CISPR, was set up under the aegis of the IEC in 1934 and comprises representatives of the IEC and other international organizations including the European Broadcasting Union, and International Conference on Large High-voltage Networks, which are interested in the suppression of radio interference to sound radio and television programmes. The UK holds the secretariat.

CISPR deals with such questions as limits for interference, methods of measurement of interference, types of interference, and impact of safety regulations on interference suppression.

Standards and exports: dismantling the technical barriers to trade

As the dust was settling at the end of the Second World War, it was the Organization for European Economic Co-operation (OEEC), set up to administer the generous Marshall Aid plan, that was the lynch-pin in all the efforts to put Europe's trade back into peace-time harness. On the standards front ISO was being set up under the aegis of the United Nations; in Europe in 1949 OEEC set the ball rolling by establishing an advisory committee for increasing productivity through standardization. Its aim was to facilitate coordination of the work of national standards organizations in the preparation and application of standards which might help to increase productivity in OEEC member countries. Mr Binney was made a member of the committee's steering group in 1951. In 1952 this advisory committee was given a new permanent role as a sub-committee of the OEEC's Committee on Productivity and Applied Research. Arising out of this, the European Productivity Agency was set up in 1953 and the OEEC sub-committee, with Mr Binney as chairman, acted as an advisory body to it, reviewing the Agency's activities in the sphere of standardization. It suggested that the agency should concentrate on work preliminary to standards of particular economic importance, or concerned with the application of standardization and simplification techniques and education services.

In the first of these two fields, the EPA initiated a project on modular coordination in building. BSI had issued a report on this subject in 1951, which indicated that a good deal of further study was needed before decisions could be made on a standard 'module' (Chapter 4). BSI and the Building Research Station undertook studies and these were combined with the EPA project for which the UK had the technical secretariat. A report of the studies made in all the OEEC countries was published by the UK in 1956. The report was followed by a programme of modular construction of buildings, BSI's test centre at Hemel Hempstead being one of the UK examples (Chapter 13).

In the second area of activity, an EPA programme of seminars on simplification in various European countries was organized in 1955 and 1956, under the EPA consultant, Professor H W Martin (Chapter 6).

The Economic Commission for Europe, an intergovernmental organization, was formed shortly after the war, also to handle economic reconstruction, and arising mainly from its work on inland transport, has concerned itself with intergovernmental agreement on regulations and in particular on standards for vehicles, components and accessories.

The Export Panel

Recognizing the vital part that exports were to play in the post-war world, BSI in 1947 set up an Overseas Standards Advisory Committee to advise on the relationship of British Standards to export trade. This was replaced in 1954 with a small Export Panel. The first chairman was Sir Ernest Goodale of the textile industry. Subsequent chairmen were Sir Cecil Weir and Sir Charles Wheeler. The panel kept under review all BSI activities affecting exports, in particular stressing the vital need for the UK's full participation and initiative in ISO and IEC committees. Three major policy statements on this subject were issued by the panel to British industry, in 1956, 1958, and 1966.

The Treaty of Rome and after

In the mid-1950s Western Europe, including the UK, was coming closer to the concept of a free trade area. In 1957 the BSI Export Panel drew attention to the importance of unification of standards for the time when this idea should eventually become reality. The six who in 1951 had set up the successful European Coal and Steel Community – France, Germany, Italy and the Benelux countries – took their far-sighted idea several steps further when, by the Treaty of Rome, they formed the European Economic Community, the Common Market. Soon afterwards the seven countries outside the Market – Sweden, Norway, Denmark, Switzerland, Austria, Portugal and the UK – joined together in the European Free Trade Association. The danger of rival ' group ' standards quickly became apparent. *The Times* noted in June 1960: ' Differing standards once they are set have proved far more difficult to unify later than have tariff requirements.' Two men who clearly saw the threat were the heads of the British and German standards organizations – Mr Binney and Professor A Zinzen. Largely as a result of their statesmanlike attitude, a meeting of the west European standards bodies at Zurich in June 1960 agreed that the Six and Seven should actively collaborate in unification of their standards within the wider ISO framework.

The practical outcome of this important decision was the creation of CEN (European Standards Coordinating Committee) and CENEL, its counterpart in the electrotechnical field. Their aim was not to duplicate or by-pass ISO or IEC, but to secure identical implementation of ISO and IEC agreements and to make use of the work done on standards by European manufacturing organizations.

In the electrical field the Common Market countries had already made considerable progress on a number of subjects by 1960 and countries outside the Common Market had to agree at the first meeting of CENEL in October 1960 that in these cases the discussion would not be reopened. On other subjects non-EEC countries took their full part and BSI was made responsible for the secretariat of groups on flameproof enclosures and tungsten filament lamps. Nevertheless there was for some time in CENEL a conflict between the interests of EEC countries, anxious to reach their own agreements, and those of the other countries.

The first meeting of CEN was held in March 1961, to agree on rules of procedure and at its second meeting in January 1962 it agreed a programme of work. Some projects were concerned with the application of existing ISO Recommendations or drafts, others with new subjects of special interest to western Europe on which no ISO work was being done, (e.g. pipe couplings for tanker vehicles), or in which special aspects of work being done in ISO needed study. A considerable programme in the building field was envisaged.

A review in October 1964 of progress in CEN working groups did not show extensive progress – resources in all the West European countries were strained both by the needs of national work and the demands of ISO – though results were beginning to show in agreement on a number of tests for petroleum products and pipe couplings for tanker vehicles; a working group on gas cylinders had made progress, though some major aspects had to await ISO work and the UK was itself held up by the long deliberations of a Government committee on this subject. At the October 1964 meeting of CEN the importance of dealing quickly with items where political considerations in EEC and EFTA demanded agreement was underlined.

This referred mainly to standards forming the basis of national regulations; the EEC was engaged on directives designed to align such regulations and the EFTA council had recently endorsed the recommendation of a working group that the EFTA countries should consult one another when introducing new regulations and should cooperate to achieve the objective of basing national regulations on international agreement.

As approval procedures were in many cases linked with regulations, a further move in CEN was to study reciprocal approval procedures in a new working group established in 1965, which took air receivers as a pilot example.

From this time efforts were made in CEN, and also in CENEL, in which the UK played a leading part, to bring the official bodies of EEC and EFTA into closer association with standards work so that the right priorities could be adopted and standards agreed for western Europe as a whole could be used as the basis of regulations in both groups of countries.

It was becoming increasingly necessary that the machinery of European harmonization should go into a higher gear. To speed up work in CEN on priority items a system was introduced in 1966, on the proposal of the Netherlands, that the standards organizations of the UK, France and Germany should take the lead in preparing initial drafts which, it was hoped, would then be acceptable to other CEN members. On this basis ' CENTRI ' groups were set up to look at a number of items including fire extinguishers, preferred metric sizes for round, square and hexagonal bars, and flanges.

At the same time, at government level, and in close association with the standards organizations, a further move was the setting up by the three countries of a ' Tripartite Committee ' to stimulate action in western Europe (in relation to activities both in the CENEL and CEN fields) and to advise on priority subjects, particularly those bearing on statutory regulations and compulsory approval requirements.

Action in the EFTA governments and standards bodies had been encouraged by a formal statement by the EFTA Council of Ministers in May 1966 recommending industries, government departments and national standards organizations to secure agreement on international standards and to accept the results of international agreement without deviation in national standards. Public purchasing departments were also urged to take full account of such agreements in their purchasing policies. The UK Government and BSI were strongly behind this statement and the policy was urged on BSI's committees with some effect. BSI also took steps to simplify the consultation procedures for conversion of international agreements into British Standards.

In its efforts to secure action on harmonized standards as the basis of regulations – and also where Governments had a particular interest as purchasers – the Tripartite Committee initiated studies of pressure vessel regulations, packaging quantities in retail trade and on arrangements for a European system of approval of electronic parts on the lines of the Burghard scheme in the UK (page 82) – on which initial discussions with other Western European countries had already been taking place. Following the Tripartite discussions on Burghard at which the general lines for a European scheme were developed, a request was made to CENEL to take over the scheme; this was agreed and the rules for a scheme under CENEL's aegis have been published.

In 1968 the efforts to ensure that the EEC (and EFTA) authorities made harmonized standards the basis of their technical directives were finally successful and it became the declared policy of both to legislate ' by reference to standards ' whenever these were available. This, and the publication of the EEC's harmonization of technical legislation programme in March 1968, gave new urgency to the west European standards organizations to accelerate agreement. This move resulted in an agreement to publish European Standards when certain clearly prescribed levels of support were reached. CEN members have also undertaken to promote and strengthen participation of their own governmental authorities in CEN work which is relevant to statutory regulations.

On the reciprocal approvals front, CEN has also since 1965 studied the possibility of having reciprocal approval schemes between its members, with a related certification mark. A corporate body known as ' CENCER ' has been set up. It owns a CEN certification mark and has drawn up rules for type testing and continuous surveillance of manufacturers' quality control. There remains the choice of suitable controls for certification and the appointment of competent independent testing and inspection organization. The importance of this venture is the service that it offers in obviating non-tariff barriers.

Another useful aid to trade is the E scheme introduced by the Economic Commission for Europe in 1958 and based on type testing only. It allows for reciprocal approval of car components such as headlamps, reflectors and so forth. This is a Government operated service but BSI's Hemel Hempstead Centre has been established as the prescribed Testing Station for testing under Government regulations.

Reciprocal approvals – a model scheme opens the way for electrical exports to Canada

Over the years, countries have worked together in the operation of bilateral and multilateral approval arrangements to ease the flow of trade. One of the best-known to British manufacturers has long been the Canadian-British scheme by which electrical and other equipment intended for sale to Canada can be approved before shipment to ensure that it measures up to Canadian regulations. After the 1939–45 war when British firms were seeking to increase their share of Canadian trade it soon became apparent that the Canadian Electrical Code and its associated strictly observed approvals system would provide a serious stumbling block, demanding as it did techniques which differed substantially from British practice: for example, insistence on double insulation and preclusion of three-core cables and three-pin plugs. Shipments of goods were found on arrival in Canada to require considerable modification, sometimes impossible to achieve. The position was specially acute in the case of bulky equipment such as machine tools.

A mission organized by BSI and led by Mr D Maxwell Buist was sent to Toronto in 1949 for discussions with the Canadian Standards Association and as a result of these negotiations it was agreed that an approvals agency be set up in Britain, to be operated by BSI on behalf of the CSA. When BSI set up its Hemel Hempstead Test Centre the scheme was operated from there, and an electrical laboratory was set up for carrying out the necessary test work involving the installation of generating equipment which reproduces Canadian voltages and frequencies.

Sir Roger Duncalfe, then President of ISO, points out to the
Canadian High Commissioner and Mr Binney a feature on a
mural presented by the Canadian Standards Association to
mark the opening of the BSI Test Centre at Hemel Hempstead
in June 1959 (see page 85)

[*To face p.* 66

Digests for the Technical Help to Exporters service being
printed at Hemel Hempstead (see page 87). The Hemel
Hempstead printing facilities are now being used for
publication of British Standards

Today the agency's list of approved equipment contains the names of 600 British firms and many thousands of separate products – from domestic appliances to heavy electrical equipment; printing presses, bottle-washing machines and hydro-extractors figure beside kettles, coffee percolators and washing machines, accounting for a great proportion of the many millions of pounds worth of electrical equipment now exported annually to Canada. The setting up of this BSI/CSA Agency was, and still is, a model of co-operation between standards bodies in different countries and has led to many other arrangements being made whereby the CSA and BSI sub-contract both testing and inspection work to each other to affect economy of operation and increased efficiency.

With the growth of industry in Canada the approvals service has become two-way, a number of Canadian manufacturers asking for their goods to be CSA-approved before shipment to Britain to ensure that they will comply with British usage. Approval arrangements have also been made between Britain and other Commonwealth countries – covering, for example, the export by British firms of flameproof equipment to Australian Standards.

Developments in Europe – the CEE

In Europe the International Commission on Rules for the Approval of Electrical Equipment – the CEE (page 59) – has since 1946 strived to achieve common approval specification for household electrical appliances. Associated with this work is the CEE's ' Certification Body ' which in the past few years has introduced a testing scheme governing products in this class of goods among its nineteen member-countries. Under this scheme, agreement has first to be reached in a CEE technical committee on an appropriate specification – for portable tools, for example – and the resultant CEE specification is then used as a basis, preferably without variation, for the relevant national standards. This provided an agreed basis for reciprocal testing. Under the scheme, equipment made to a CEE specification and tested and certified in the country of origin and in one other country was then accepted in all other member-countries of the CEE. A UK proposal to cut out the requirement for certification in the second country, was accepted in May 1969. The application of a CEE mark to denote approval of equipment under the reciprocal testing scheme, and involving continuous surveillance of protection, has been under consideration for several years but agreement has not so far been reached.

Technical help to exporters

While the back-room work continues to achieve harmonization of standards and of approvals schemes, firms involved in export trade need immediate technical assistance in coping with the everyday problems that arise. Of enormous help has been the Technical Help to Exporters service set up by BSI at its Hemel Hempstead centre in 1966. The service performs a dual function: it provides information to exporters as to differing legal, approval and standards requirements with which they may have to comply, and it helps them in securing the necessary approval of their products in advance of shipment. Five hundred firms joined the scheme in its first three weeks of operation, anxious to take advantage of technical digests giving complete information on specific categories of equipment country by country. In view of the problems surrounding the export of electrical goods in particular, it was decided to give first priority to information on export of all types of electrical materials to the countries of Western Europe. Help in resolving specific problems has always been readily obtainable.

What are the regulations governing mains-driven motors for recording instruments in Germany? What do the Australians specify for the internal wiring of machine control gear? What are the regulations/standards governing steel joists in Hungary, cryogenic vessel design in the Netherlands, blast furnace cement in France? Such questions as these are all grist to the Technical-help service. One useful advance in this area was an American decision in 1970 to accept the reports of certain foreign test houses – such as the BSI Hemel Hempstead centre – in respect of motor vehicle equipment exported to the United States.

Another important development in 1970 of BSI's services to UK companies exporting to the United States was an agreement with the American Underwriters' Laboratories whereby BSI would follow up inspection of UK goods which have been approved by the American body. For many years products such as electrical appliances and fire-fighting apparatus have required the Laboratories' approval before they can be sold in the US; now the frequent follow-up checks on the production line are BSI's responsibility. Similar arrangements have recently been made with the Canadian Underwriters.

BSI engineers engaged on the Technical Help to Exporters service have to date visited some 40 countries and have written nearly 200 technical digests; 1200 firms are members of the scheme which has played a significant part in helping to boost export trade during the past few years.

At BSI's headquarters London much has also gone on to make easier the task of exporters. A central enquiries section was established in 1969 to help route enquiries to the right department; the standards publications of over 50 countries were grouped according to subject matter to make information retrieval more straightforward; translation of overseas standards into English was greatly accelerated.

Standards for the shopper

The rise of the consumer protection movement

Ultimately the object of producing standards must be to improve in some way the goods and services which are available to the nation as a whole. Standards for steel may be written by and for the steel industry and its commercial customers – but at the end of the production line it is Mr Smith who benefits in being able to buy a cheaper, more reliable car or dishwasher. So this chapter looks away from the vital mainstream of standards for industry to concentrate on the efforts made from the end of the 'thirties to develop standards for goods bought over the counter in retail shops. It takes in the growth of interest in Britain – and the rest of the world – during the nineteen-fifties and 'sixties in the subject of protection for the consumer: a phenomenon which reflected the changing social scene and was in part a revolt against high-powered advertising.

Start of consumer standards

A few, not very well known, standards for consumer goods were published in the 'thirties. On the outbreak of war in 1939 the Retail Trading–Standards Association, a body concerned with maintenance of standards of good practice, invited BSI to join with it in establishing consumer standards on a more effective basis. Clearly, at a time of shortage and rationing the best possible use had to be made of scarce materials and so the Distributive Industry (Standards) Committee came into being in May 1940, a development that was somewhat overshadowed by the darkening events of war. The new committee took as its first task the question of clothes sizing as a measure of economy with BSI asked to lay down sizes for a wide range of women's and children's clothing.

There followed an invitation from the Board of Trade for BSI to prepare the specifications needed for the new ' Utility ' schemes by which throughout the war and after the whole population were assured of getting reasonably well made – if not particularly glamorous – dresses, underwear, blouses, boots and shoes and household textiles. BSI thus became significantly involved in a whole new range of interests.

With the end of the war in 1945 there came an opportunity for further progress. The National Council of Women set up an Advisory Committee on Consumer Goods, one of its main tasks being to review what was available, and what was required, in the way of standards. Some useful new standards were being developed – notably the standard for fillings used in mattresses, pillows and upholstery – and standards for various items of women's and children's wear.

Post-Utility – the big break-through

In 1951 a new Conservative government set up the Douglas Committee to consider the future of the Utility schemes and accepted its recommendations that Utility should be abandoned in favour of quality assurances worked out between the industries concerned and BSI. These industries, glad to be free of the tight control exerted for so long by the Utility schemes, were not slow in giving their promise to co-operate. But from the start, difficulties were encountered with the textile and clothing trades, whose zeal for control waned as the Utility schemes receded. In 1953 BSI faced up to the realities of this situation and announced that future work would be quite divorced from Utility-type specifications and would concentrate on informative labelling and performance testing, with detailed constructional specifications confined to a few products – for example specialized cloths such as tickings.

Much more successful were the standards for furniture and bedding. A standard for domestic bedding was published in 1952 and the first parts of a standard for furniture issued a year later.

The Women's Advisory Committee is set up

Throughout the country there were signs of increasing interest in the welfare of the ' consumer '. It was a subject often touched on in Parliament and the newspapers, notably the women's pages, were full of it. The Cunliffe Report (page 18) was reflecting this interest when it said that representation of the domestic consumer in the preparation of standards was a matter demanding serious attention.

The main reason for this concern was a growing awareness among thoughtful people that the post-war world was going to be a very different place and certainly not least in the choice of goods put before shoppers. At this time such organizations as the Consumer Council, Consumers' Association and the National Federation of Consumer Groups did not of course exist.

In 1951, in direct response to these pressures, BSI set up its Women's Advisory Committee, a much more representative successor to the *ad hoc* standards committee formed at the end of the war by the National Council of Women and, as it was to turn out, one of the most enduring moves in the story of consumer protection. When the committee first met round the table in 1951 it numbered 18 who together represented the views of one-and-a-half million women – the membership of the most important women's organization in the country, including the National Council of Women, Women's

Voluntary Services, Townswomen's Guilds, Federation of Soroptomists Clubs, and Women's Institutes. Today the WAC represents 31 organizations with a total of over four million members.

Under its extremely able secretary for 16 years, Mrs Margaret Thompson, the WAC's function was to bring women's practical experience to bear on any aspect of BSI work affecting the ordinary consumer – to effect a real degree of collaboration between manufacturer, retailer and shopper. The job of the WAC then as now was to tell the members of the BSI technical committees the sort of standards their members – and women in general – wanted, by making proposals for new standards and seeking to get existing drafts amended to suit women's needs. Over the years there have been innumerable examples of standards being improved as a result of cogent WAC comment, whether in regard to a more convenient height for kitchen working surfaces, greater safety of spin driers, or electric blankets less likely to result in fire or shock.

But from the start the WAC was also looking beyond the immediate present to the barren wastes where there were as yet no consumer standards at all – to such things as quality of school uniforms, the wearability of lino, or the provision of special furniture for the elderly or infirm.

The WAC has a second important function – to make known by publicity to members of its constituent organizations as much as possible about consumer standards and safeguards. Its publication *Consumer Report* which started in 1954 has a circulation of 26 000 copies. It has organized conferences throughout the country for women's organizations, teachers, students and specialist groups to explain different aspects of BSI's work for consumers. It provides leaflets for shoppers, and comprehensive lecture material for teachers.

Development of consumer standards and Kitemarking

Throughout the 'fifties there came a steady flow of consumer standards and Kitemark schemes, resulting in BSI becoming increasingly identified in the public mind as a champion of the ordinary shopper.

A series of standards for domestic furniture was completed in 1953 and associated with them, a revolutionary set of performance tests to simulate rough wear. The tests, demonstrated at the furniture exhibition at Earls Court in 1954, captured the imagination of both furniture manufacturers and the general public to such an extent that BSI was able to report that year that no fewer than 300 firms had been awarded licences to Kitemark their furniture – well over half the industry's output. At the same time the bedding industry was able to make the claim that 85 per cent of its output was now Kitemarked and that shoddy mattresses were a thing of the past. The pressure cooker, so much a part of the kitchen scene in the 'fifties, was another product extensively sold under the Kitemark label.

Some of the Kitemark schemes – furniture for example – did not live up to their first fine promise, with quite a few firms dropping out as time went by. But those where personal safety was a significant factor have been more enduring. In 1953 a standard for motor-cyclists' helmets was issued, and this standard, backed by skilfully devised laboratory tests to simulate the kind of treatment a helmet might receive in an accident, was a precursor of many more aimed at safeguarding people at work and at play such as the standards and Kitemark schemes for car rearlights and reflectors and seat belts.

Standards are essentially voluntary agreements and do not normally carry legal endorsement. But reflecting the feeling that consumers needed more protection, a few standards were given the force of law, usually by quoting their basic requirements in legislation, and the Kitemark can give the user a ready and visible assurance.

Before the war and for years afterwards, women had complained about the multiplicity of sizing systems with which they were faced when they were buying clothes. In February 1953 BSI brought together manufacturers, wholesalers and retailers to sort out the prevailing chaos and replace it with an agreed system. A year later came agreement of the trade to a schedule of sizes for both women's outerwear and children's wear. In fact this did not prove satisfactory because people are not of standard shapes! It has been replaced by a standard for size marking, enabling different manufacturers to cater for different figure shapes, while presenting in a standard form the principal measurements of the figure the garment is designed to fit.

The Consumer Advisory Council and Shopper's Guide

In response to ever-growing interest in the subject of protection and guidance for shoppers, BSI took its headiest plunge yet into the deeps of consumer protection. For in January 1955 it set up its Advisory Council on Standards for Consumer Goods (later abbreviated to the Consumer Advisory Council), a body embracing a far wider range of interests than the Women's Advisory Committee – though the WAC was to be strongly represented on it – retailers, wholesalers and individuals with a strong interest in consumer matters including public relations experts and journalists. Manufacturers as such were excluded from the Council and, from its inception, the sensitiveness of some sections of industry as to the Council's activities was to prove a source of difficulty. BSI did not see any real dichotomy of interest since neither it nor the CAC regarded trading as a battle of wits in which producer and consumer were on opposing sides, but it realized the need to tread warily.

An admirable summing up of the situation was given by Mrs F H Shepherd, the secretary to the CAC, when she addressed a management conference in November 1957. 'Never before', she pointed out, ' has there been so vast or so variegated a choice of goods spread before the consumer.

Where our grandmothers, or their domestics, cooked with cast iron vessels over coal fires, their grandchildren have the choice of aluminium, stainless steel and glass, in many patterns and sizes, for use on gas, electric or oil stoves. A hundred years ago the shopper had the choice of wool, linen and cotton; today, with the advent of man-made fibres and their rapidly increasing diversification and blending, and with the evolution of new finishes for the traditional fibres, the consumer of today can have little idea of what she is buying or what can be expected of it. The average shop assistant is all too often equally at sea.

' If the waste of bad buying is to be avoided, the consumer needs, at the point of sale, more and better information which she understands. . . . The cult of the label as an advertising medium is far advanced, but it will soon defeat its object unless it provides factual and reliable information.'

The Consumer Advisory Council started its career by sending to 20 000 members of the general public, a questionnaire asking them to name the goods on which they most wanted information, and the kind of information they most urgently required. The replies poured in, with the majority of votes going in favour of children's clothing, fabrics for dresses and underwear, blankets, furnishing fabrics and carpets. This was one aspect of the CAC's work – to ascertain the needs of consumers and to make them known to BSI with a view to satisfying them through provision of standards. The other aspect was to provide shoppers with an information service and to accumulate evidence as to goods and defects which caused people most difficulty. Perhaps, most important of all, the CAC soon became a focal point for national publicity about consumer standards.

In 1956 with the aid of a Government grant of £10 000, the CAC decided to enroll individual consumers through its 'Associates' scheme, and there was an immediate response by tens of thousands of men and women, demonstrating the desire of people for some regular means of receiving advice on their shopping problems. A year later this need was more fully met with the issue of something quite new in British journalism – the magazine *Shopper's Guide* – which under the editorship of an already well known crusader in this field, journalist Elizabeth Gundrey, pioneered in comparative reports on different kinds of consumer products.

Only weeks after the first appearance of *Shopper's Guide* came what was to prove a formidable rival, the Consumers' Association magazine *Which?*, a publication less inhibited than *Shopper's Guide* and whose ' best buy' summing-up at the end of each of its reports won public approval. Meanwhile, *Shopper's Guide* itself did not pull its punches and its critical reports of products sometimes made by BSI's industrial members brought home to manufacturers that the day of the complete seller's market had come to an end. For BSI there were of course some embarrassments in all this and by the end of 1958 there were many – not least members of the CAC itself – who felt that consumer protection needed to be more than just an offshoot of an organization whose primary task was something quite different. A Government announcement in July 1959 that it was to set up a committee of enquiry into the whole subject was greatly welcomed, and BSI asked that special attention be given to the position of the CAC.

While the committee, under the chairmanship of Mr J T Molony, Q.C., was hearing evidence, the CAC's work continued at a high and successful level. Consumer ' brains trusts ' in provincial cities encouraged the setting up of local consumer groups throughout the country. Consumer protection really had arrived, it was felt, when BBC television began its ' Choice ' programme, with Richard Dimbleby presenting selected reports from both *Shopper's Guide* and *Which ?*

An interim report by the Molony Committee in April 1960 recommended that the sale of appliances with potential dangers to the user should be controlled by legislation. This resulted a few months later in the publication of the Consumer Protection Bill, designed to give the necessary powers to the Home Secretary. Under this Act, regulations have been introduced over the years to prohibit the sale of a number of products not complying with the appropriate British Standard.

When the Committee made its major report in the summer of 1962 BSI commented: ' Though the report does not point the way to a revolution in consumer protection methods even its modest positive proposals would hardly have been possible to contemplate if public interest had not been aroused in the consumer and his problems – and in this the BSI's Consumer Advisory Council has played a large part as has also the Women's Advisory Committee.'

The Committee's major recommendation was that a national Consumer Council should be set up, but that it should be excluded from undertaking comparative testing or operating a complaints service. On the question of standards the Committee fully endorsed BSI's traditional role but recommended a tougher line in publishing consumer standards, even to the extent of issuing a standard against substantial manufacturing opinion. In future, it said, BSI should concentrate on informative labelling and standard sizing. Adequate information was the first line of defence in any system of consumer protection, and power should be sought from Parliament to impose compulsory labelling requirements by regulation if a case for such action was established, stated the report.

BSI's reaction to some aspects of the report – especially its underrating of the value of quality standards as a safeguard to purchases – was not one of pleasure. The committee came down strongly against BSI practice of making Kitemarking a compulsory feature of certain safety standards. It was considered that a case against a manufacturer on this count would not stand up in a court of law. As a result, the compulsory Kitemark clause was phased out, although a strong recommendation that the Kitemark should be used in appropriate cases was made instead. The winding up of the CAC and

Richard Dimbleby introducing the BBC television series
'Choice'. The series, held in cooperation with the Consumers'
Association and the BSI Consumer Advisory Council,
featured some of the reports published in *Shopper's Guide*
(see page 74)
(*Photo courtesy BBC*)

To face p. 74

Mrs A Stanley, a past chairman of WAC (left), Mr G B R
Feilden, Mrs C Davis, chairman of WAC, and Mrs M
Thompson, former secretary of WAC, examine British
Standard toys on display at the 1971 WAC conference

Shopper's Guide which resulted, did however relieve BSI of some embarrassing problems which had promised to grow rather than diminish.

Efforts were made to provide for the continuing publication of *Shopper's Guide* by another body – a publishing enterprise with a trust to secure complete independence was envisaged. In the event, however, this did not work out and the magazine ceased publication. When the new national Consumer Council came into being the CAC was disbanded after an extremely eventful seven-year history.

After the CAC's demise, relations with the Consumers' Association, publishers of *Which?*, were strengthened with greater BSI co-operation in relation to tests used in *Which?* assessment of goods, and with CA being given representation on BSI technical committees dealing with subjects on which it had gained experience.

Now the Women's Advisory Committee was left as the direct link between BSI and consumers, with the WAC chairman appointed as a member of the new Consumer Council, and the Council itself nominating representatives to BSI technical committees.

With so much emphasis being placed on the need for informative labelling, the Consumer Council was quick to concentrate on this aspect and at the end of 1964 introduced its ' Teltag ' labelling scheme. For this scheme BSI undertook to prepare the standard tests on which the information to be given on a Teltag would be based – for example, construction, weight, capacity, durability, resistance to light or water. Among the first subjects chosen for Teltags were: electrical appliances, clothing, blankets, carpets and hardware.

The Consumer Council was active in many other directions too – in calling conferences to highlight particular problems, in working towards new and improved consumer legislation and in producing educational material for schools and the public generally.

The Council was, alas, to have no longer a life than the CAC, the Government axe falling on its grant resulting in its closure in March 1971. Under its first and only director, Dame Elizabeth Ackroyd, it accomplished a great deal in its seven years' existence, drawing attention to, and helping to stop, abuses such as ' switch ' selling and the sale of sub-standard goods in many fields. The pity was that by 1971 there was no independent organization willing to take over the financial and other responsibilities of running such an organization. The splendid Joint Committee on Consumer Complaints which had been started by the CAC, following a suggestion by Lord Snowdon, and expanded by the Consumer Council, is still continuing under the aegis of the Institute of Weights and Measures Administration. The National Council of Social Service has set up a committee of interested parties to see what other Consumer Council activities can be continued on a voluntary basis.

Standards and consumer safety

Much of BSI's effort in the consumer field has always been concentrated on safety. A particular landmark was the publication in April 1957 of a report on flammability of fabrics – the result of an investigation into the grisly burning accidents which occurred all too frequently, involving death and serious injuries of small children and old people. It showed that practically all the lighter types of fabric were equally susceptible to flame and advocated greater efforts to have all fires and heaters properly guarded, the need to label the safer materials (such as certain synthetics or fabrics treated chemically to make them flame-resistant), and the desirability of dressing children in close-fitting clothes such as pyjamas, not flowing ones like nightdresses which catch fire easily.

The report – and an explanatory leaflet of which tens of thousands were distributed in a few weeks – shocked government, industry and other interests into an awareness of the seriousness of the problem. There followed the publication of a new standard for the durable flame–resistance of fabrics and this was adopted in regulations laid down by the Government in 1964 which prohibited the sale of children's nightdresses made of a fabric not in accordance with the standard. Subsequent legislation has made it compulsory for women's nightdresses not made of a safe material to carry a warning label.

The British Standards for fireguards have also made a notable contribution to safety, although the high quality of these guards has unfortunately tended to restrict sales. These guards have either to fit securely to the fireplace surround or be of a shape and size to surround the fire completely, so keeping young children well away from flames.

Although the standards are revised as necessary to keep them up-to-date, occasionally such revision is behindhand; such a case was the standard for paraffin oil heaters. The requirements in the original standard were adequate for the convector heaters which were the only sort made at the time. It did not provide adequately for the radiant ' drip-feed ' variety that became popular in the late 1950s.

The standard was revised in 1963 and its effect strengthened by corresponding government regulations; BSI set up a new Advisory Committee on Personal Safety with the aim of alerting all committees concerned with safety to possible future hazards. The object was to bring out as early as possible the need for new or improved standards and to identify potentially dangerous trends from evidence of accident statistics. The committee, which for many years met with BSI's director general as chairman, brought together representatives of such bodies as the Royal Society for the Prevention of Accidents, the Home Office, Ministry of Health, and Fire Protection Association. It has considered the safety of electric blankets, perambulators, toys and many other items.

Further development of consumer standards in the 'sixties

In 1960 came a standard which was destined more than any other to put the Kitemark on the map – this was the standard for car seat belts which, with the addition a year or so later of its highly dramatic ' dynamic ' test for crash-testing sample belts, was incorporated eventually into Government regulations, making the fitting of approved seat belts compulsory in all new vehicles.

Another major advance was the introduction of a comprehensive series of standards for the newly-formed British Electrical Approvals Board under which domestic electrical appliances, from kettles to vacuum cleaners, heaters to hair-driers, dish-washers to waste-disposal units, were to be tested and approved for safety, and to be labelled with the BEAB Kitemark.

A code of safety precautions for toys, to which UK manufacturers promised to adhere, recommended that fillings and coatings should be clean and non-toxic, that there should be no sharp edges, that eyes and joints should be securely attached, that clockwork mechanisms should be enclosed, and that toys should not be made from celluloid or other highly combustible materials.

A code for informative labelling of carpets, published in 1963, required manufacturers to state the size, type, fibre content and instructions for cleaning. In 1965 came the textile care labelling code which provided uniform instructions through words and symbols for washing, ironing and dry-cleaning. By 1966 BSI was collaborating with the Consumer Council on its ' Teltag ' schemes and was developing the necessary performance tests.

Much remains to be done. Today the WAC has representatives on some 160 technical committees preparing British Standards for consumer goods and it has a panel of some 400 correspondent members spread throughout the UK who make local enquiries and fill in questionnaires. It continues to point out that it wants to see many more goods in the shops with a BS number on them and to urge that manufacturers should be taking much more interest in the Kitemark. Unfortunately the Trade Descriptions Act has provided manufacturers with a good excuse to give as little information as possible! The WAC is deeply involved with consumer education on a broad front – notably in respect of the change to metric – and its views increasingly are sought by Government and by a host of organizations at home and overseas. For the rest it seems clear that BSI's programme of work on consumer standards will remain a long one for many years to come.

12

Buying with assurance : the Kitemark and other certification schemes

The idea of offering buyers a guarantee of quality is not new. Craftsmen in the Middle Ages, anxious to protect their good name from any accusation of shoddy workmanship, evolved standards of performance and instituted marking systems to denote quality, the hallmarks for gold and silver still being with us.

So it was that almost as soon as the first British Standards were being prepared at the beginning of the century, the need was felt for some form of mark to indicate to buyers that goods were 'up to standard'. The British Standard mark – to become familiarly known later as the Kitemark – was first registered as a trade mark for tramway rails back in 1903. It had little impact until the Trade Marks Act was amended in 1919 to allow for certification of manufactured articles on the evidence of tests carried out on a *sample* of the total production, not inspection of *every* item as had hitherto been the case. Two years later the British Engineering Standards Association set up a Mark Committee to work out a procedure under which it could grant licences to firms wishing to use the mark to certify compliance with a standard. In 1922 registration of the mark was accepted for all classes of product and the first licence for its use by a manufacturer was issued in 1926, to the General Electric Company.

In 1930 the term 'British Standard' was also registered as a trade mark, not only to serve as an identification that a product complied with a standard, but also to prevent misuse of the words by manufacturers.

For more than 40 years, therefore, firms producing to a standard have been able to claim compliance by marking a product with its BS number. For over 20 years many standards have indeed contained a positive requirement for the marking of the product with its BS number – together with the maker's name and other basic information about the product. Marking in this way implies a unilateral claim by the manufacturer himself and any misdescription is a matter to be dealt with in law. BSI would of course take up with a manufacturer any case of misuse brought to its attention; on occasions evidence has been passed to the Department of Trade and Industry which has prosecuted firms for misdescribing goods as complying with British Standards when they did not. (The Trade Descriptions Act, 1968, now provides for local authority Weights and Measures departments to bring such cases).

Where a British Standard is referred to or is incorporated in legislation –

for example, in the regulations for motor cyclists' helmets – any offence is dealt with under those regulations.

Development of Kitemarking

A major landmark in the history of certification marking came in 1937 when owners of motor vehicles were legally obliged to fit safety glass in their windscreens. A British Standard for safety glass was published and the principal manufacturers marked screens with the ' Kite ' to show conformity with the standard. After the war came another large-scale scheme administered jointly by BSI and the British Plastics Federation, for certification of plastics materials and products. In 1939 there had been 171 firms licensed to use the Kitemark. By 1950 the number had risen to 540 and it was to rise sharply during the following years as a great many new schemes were introduced.

Two of the most important related to the furniture and bedding standards, adopted by these industries following the withdrawal of the war-time and post-war compulsory Utility schemes. At their peak, in 1955, some 300 firms were selling Kitemarked furniture (as was mentioned in the preceeding chapter), and some 85 per cent of bedding was Kitemarked. The bedding industry could, and still does, claim with justification that adoption of British Standards and the Kitemark signalled the end of the shoddy mattress. Other notable Kitemark schemes for consumer goods were those for electric blankets – unquestionably putting this new product right on the map and contributing to a leap in sales of from 150 000 to one-and-a-half million in ten years; oil heaters – a major safety scheme run in collaboration with the Oil Appliance Manufacturers Association and subsequently built into government regulations; numerous items of safety equipment including helmets for motor cyclists and horse riders, and life-jackets for yachtsmen; seat belts for cars – which probably more than any other scheme has made the Kitemark familiar to vast numbers of the public (five million Kitemarked belts being produced in 1967/8). Seat belts are one of those products on which the presence of the Kitemark is *prima facie* evidence of compliance with legislation.

The mark is today used in connection with about 250 British Standards covering an estimated annual production value of around £500 million; some 90 per cent of the products covered is in the area of industrial equipment.

On the industrial and commercial side, of special value to large-scale purchasers, have been the Kitemark schemes for zinc-alloy diecastings, industrial eye protectors, copper tubes, flameproof electric motors, road traffic signs, PVC pipes, portable tools, street lanterns, school furniture, and a great many more.

Certification by BSI through its various schemes provides an independent assurance of the compliance of a manufacturer's products with the relevant standard. A system of supervision and control by BSI is drawn up for each scheme. After a representative production sample is found to be fully in accord with the standard, continuing compliance is ensured by regular audit of the firm's quality control arrangements (which have to be approved before a licence is issued), and the testing by BSI or agreed agents of a proportion of production – often determined by statistical methods. BSI Kitemark schemes have had the side effect of vastly improving quality control in firms – indeed they have in many cases been the way of introducing it.

The problem of the isolated defective item which is always liable to occur in large-scale production, even with excellent quality control methods, has caused BSI concern from time to time. While the 'Downland Bedding' case in 1958 made it clear that the law did not exonerate the manufacturers from the falsity of the claim of compliance in respect of the defective item itself, it clearly took the evidence of the interests of the certification scheme into account when passing sentence. There is a real danger that the provisions of the Trade Descriptions Act may discourage the giving of information including the Kitemark itself, since insufficient allowance is made for the statistical realities of mass-production methods. These pose a dilemma for the legislator where individual purchases may include the exceptional defective item.

During the 'fifties and 'sixties a number of organizations were promoting marks and seals of various kinds – some of extremely doubtful reliability; the danger was that unknowing purchasers might well confuse such seals with the Kitemark. When BSI gave evidence to the Molony Committee on Consumer Protection (page 74) it was specially anxious therefore to define its authority in regard to certification marking. The view was strongly put that any marking scheme, to be effective and enjoy public confidence, must be impartially administered by an acknowledged, independent body, must use published standards as its criteria of approval, have adequate control at the source of production and ensure a consistent standard of quality.

It had been BSI policy for some years to encourage standards committees to require certification marking as a compulsory part of certain standards – notably those affecting safety and health. However, the Molony Committee did not consider that such a requirement would stand up in a court of law and, since the Board of Trade accepted this view, the policy of compulsory marking had to be abandoned in favour of a strong recommendation for Kitemarking where appropriate.

The highest number of licences was achieved in the 'sixties when some 1450 firms were Kitemarking their products. There has been a slight falling off in recent years owing to mergers and firms going out of business, lessening of interest in the big furniture scheme and growing emphasis in the consumer field on labelling rather than certification. Publicity for the Kitemark has never really been sufficient to create a strong demand for it. On the whole manufacturers have not been over-enthusiastic, not prepared

The issue of BS 3595 for lifejackets in 1963 was followed by the adoption of its more stringent requirements in Government legislation (despite earlier opposition to the BSI proposals!)

(Above) Children demonstrate a version of the lifejacket designed for young people, and approved under the Kitemark scheme

(Below) At a meeting of the ISO technical subcommittee on lifejackets held in 1964, Dr J E Gabb, chairman of the corresponding BSI committee, 'fits' a delegate with an approved lifejacket.

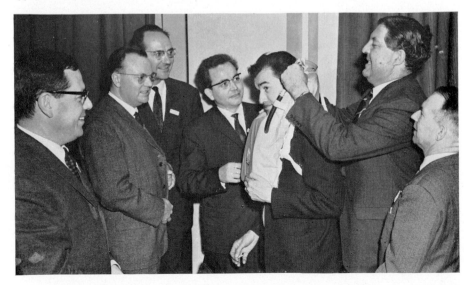

[To face p. 80

Mr Ernest Marples, Minister of Transport, tries on motor-
cyclists' helmets approved under the BSI Kitemark scheme.
A nationwide road-safety campaign for the wearing of helmets was launched in August 1963

to spend large sums in advertising although in a number of cases they have established a product on the market under the banner of the Kitemark.

In 1958 the words ' Approved to British Standard ' – plus the BS number concerned – were added round the mark to make it more meaningful and in subsequent years a series of attractive Kite labels and swing tickets were designed to help buyers more readily in identifying approved products. Catalogues listing the names of firms making Kitemarked goods have also helped in this respect.

There has been a steady increase in awareness among large-scale purchasers, such as local authorities and government departments, of the value of ordering Kitemarked goods and so eliminating their own testing and inspection. BSI's Local Authority Standards Advisory Committee has helped in this, as did the endorsement of Kitemarking in a government White Paper on Public Purchasing and Industrial Efficiency issued in May 1967.

Where compliance with a standard is required by legislation, use of the Kitemark is often accepted as evidence of compliance, which provides a practical means whereby the manufacturer can satisfy himself and the law that his product complies. Thus the regulations in 1958 requiring car rear-lights and reflectors to comply with British Standards resulted in a large increase in applications for the mark. All motor cyclists' helmets are required to comply with the appropriate standard and proposed regulations will require them to carry the Kitemark.

Over the years, firms overseas – particularly in western Europe and Ireland – have also been granted licences to Kitemark their products destined for sale in the UK and there are now some 70 foreign licensees in different countries.

Joining with other bodies

For many years BSI has collaborated with trade associations of high reputation in operating Kitemark schemes on behalf of their member-firms – the British Plastics Federation and Oil Appliance Manufacturers Association, as already mentioned, and with the British Waterworks Association on joint certification of taps and other water fittings. Since 1958 the Gas Council has operated its own approval scheme for all domestic gas appliances, using British Standards as the criteria.

Proposals for a special approvals scheme for domestic electrical appliances were being intensively canvassed during 1957. At a time when safety marks were very much the ' in ' thing, manufacturers of electrical appliances desired a special mark of their own – and also wanted a British equivalent to the approvals organizations which had been set up in several European countries. One long-debated question was the part the Kitemark might play in the proposed scheme, bearing in mind that British Standards were to be employed as the criteria for approval. In fact when the new British Electrical Approvals Board met for the first time in February 1961 at British Standards House it had decided to use its own ' electrical ' Kitemark.

Under the chairmanship of Lord Citrine the Board brought together representatives of the British Electrical and Allied Manufacturers Association, the electricity supply industry, contractors, wholesalers and retailers, and of BSI. The criteria were to be a new series of standards which was to involve BSI in a crash programme of revising existing standards to include the necessary safety tests, and preparing new standards where required. Ten years after BEAB's inception, when the Women's Advisory Committee held an anniversary reception for BEAB, it was announced that its current list gave approval to nearly 1600 different appliances. Electricity Boards and reputable stores stock BEAB-approved appliances, to the great benefit of the shopping public.

BSI was also closely associated with the setting up in 1963 of a scheme for approval of cables by the British Electric Cable Testing Organization, and of a scheme set up in 1968 for approval of oil-burning equipment by the Domestic Oil Burning Equipment Testing Association.

The Burghard scheme

An approval scheme developed in Britain and now accepted as a basis for multilateral trading in western Europe has brought coherence, appreciable time-saving and independent assurance of inspection to the electronics components industry – previously overburdened with a complex variety of specifications. This Burghard scheme, as it is known, results from a report in 1965 of a government committee under the chairmanship of Rear-Admiral Burghard, which had reviewed the whole field of electronic component reliability, with particular reference to the needs of the armed services.

The scheme, based on BS 9000, provides a unified series of specifications for electronic parts for both civil and military purposes, but within this common framework allowance is nevertheless made for the speedy introduction of new component designs; it therefore achieves a flexibility seldom obtained in schemes of such tight quality control. It has, in fact, set a new pattern for standards and certification in that quality control is built into the specification and there is an integrated system to provide for the production of electronic parts of assessed quality at the levels required for different uses. The first basic specifications were published in May 1967 and the first approvals of firms – manufacturers, stockists and test houses – issued in 1969. Specifications, of which there are now over 2000, are normally based on recommendations of the International Electrotechnical Commission.

Standards and quality control

Quality assurance begins in the control exercised in the production process. If this control is right, the achievement of the quality defined in the specification is ensured and the procedures associated with the certification of a product are simplified. BSI has taken an active part in the application of quality control principles in standardization and in the development of quality assurance through the certification of products manufactured to comply with standard specifications. It is only within the past fifteen or twenty years that the significance of developing specifications on a statistical basis has been recognized. In BSI the origins of the idea go back to 1935 when BS 600, a standard which deals with the application of statistical methods to industrial standardization and quality control, was published. It was primarily the work of Professor E S Pearson, FRS. At the time of publication BS 600 was an original contribution to the science of standard-ization at the national level, and it has maintained a position of importance as a key work on the application of statistical principles ever since.

Many British Standards have been published in which the general principles of statistical theory applied to standardization are described. In addition a number of British Standards specific to particular products, and methods of testing them, define the quality of the product in terms of a statistical approach which lends itself to incorporation in quality control procedures used in the manufacture of the product.

Addressing the 1962 standards engineers' conference, Mr F Nixon, then chairman of the National Council for Quality and Reliability, claimed that although British Standards were respected all over the world, standardization was having less impact on British industry than in many competing countries – he was specially concerned with the setting up of quality standards. Vague requirements that ' a product must be free from defects ' were useless. With increasing adoption of automatic methods of manufacture it was increasingly necessary to specify tighter tolerances on the dimensional and physical characteristics of the materials used.

The following year, 1963, was ' National Productivity Year ', a 12-month campaign to boost productivity in the UK, to focus attention on every possible means by which management and workers could raise efficiency and lower costs.

A leading article in *BSI News* at this time asked: ' Does industry pay sufficient attention to the role of British Standards in quality control? There are standards which provide expert guidance. Indeed, one of the most important services that the BSI can provide during National Productivity Year is to convey the message to the country at large that efficient quality control in manufacture, aimed at the production of goods and materials complying with national standards to ensure the products are fit for their purpose, is one of the sure ways by which productivity and the standard of living will be raised.'

Quality and Reliability were in fact to have their own ' year ' in 1966–1967, in an effort to raise quality and lower costs. Lord Kings Norton, on his appointment as first president of the National Council for Quality and

Reliability, said Britain had to establish quality and reliability standards for the markets in which we aimed to succeed, and to produce consistently to these standards. In June 1967 came the publication of a *Guide to the preparation of specifications,* the work of a NCQR Committee, giving a framework of headings under which requirements could be drafted in a logical and comprehensive form – a useful guide for standards engineers and others concerned with the preparation of company standards and other procurement documents.

A report on the effects of Quality and Reliability Year issued by the BPC and NCQR at the end of 1967 brought out very clearly the key position of national standards in the country's drive for quality. The verdict was that working to clearly defined specifications pays dividends and that special requirements laid down by every large purchaser can only be wasteful.

Recent developments in certification

In September 1968 the then Ministry of Technology and the Confederation of British Industry set up the Mensforth Committee to consider the possible extension and co-ordination of quality assurance schemes, with special reference to export requirements. BSI submitted detailed evidence. The Committee issued its report in March 1971, which drew attention to the problems of quality assurance facing exporters of engineering goods and materials to other countries and it pointed out that in the main these problems arose out of the need to demonstrate, often in a particular way, that such goods or materials conform to specific requirements. Such requirements were often imposed by the foreign customer to satisfy his particular needs or meet the mandatory requirements of the country in which he resides. The report concluded that there was a case for a more rational arrangement of existing quality assurance organizations and proposed a ' federal quality organization ' to achieve this.

Meanwhile BSI, in order to guide and stimulate the future development of independent certification and quality assurance, had come up with its own answer. Its Certification Mark Committee, which had for long been responsible for all certification schemes, was replaced in 1971 by a new Quality Assurance Council, responsible under the Executive Board for the policies to be followed by BSI in the whole field of certification, approvals and marking, nationally and internationally. Much is hoped for from these developments, since quality assurance work is increasingly seen as a useful tool in helping the acceptance of products in international trade.

13

Testing to a standard: the first twelve years of the Hemel Hempstead Centre

' The establishment of this Centre confirmed BSI as a complete standards organization in the up-to-date sense of that term – that is, an organization concerned not only with specification, as such, but equally with systems for ensuring compliance with specification.' These were Mr Binney's words on the occasion of the opening of the Sherfield building in 1971, when a plaque honouring him as the founder of the Centre in 1955 was unveiled. ' By this move,' he went on, ' it was also confirmed that BSI would continue as the leading standards organization of the western world '.

An essential feature of certification to standards is that completely reliable test facilities should be available. Until the end of the 1950s BSI relied almost entirely on the services of independent – and highly respected – laboratories and test houses for this work. With the growth of certification marking it was however becoming apparent that BSI, without duplicating existing facilities, needed to have test facilities of its own. This would enable a closer control of Kitemark schemes and a continuing liaison between, on the one hand, testing staff, and, on the other, the BSI technical committees which produce the standards. Above all, such a centre reflected the needs of the British export trade.

To meet these needs, BSI opened its test centre on the new industrial estate at Hemel Hempstead in June 1959. It was a modest single-storey affair, built appropriately enough to standard modular principles, but the three acre site on which it stood was evidence that further expansion was envisaged. The original three buildings were used to house the BSI/CSA Approvals Agency, formerly housed at British Standards House, and several testing units for existing Kitemark schemes, including laboratories for the testing of motor cyclists' helmets (one of the biggest certification schemes at that time) and the cleanliness of filling materials. There was also a small electrical laboratory to undertake testing work related to the Canadian approvals scheme. It was also intended to provide facilities for organizations and manufacturers wishing to commission private test work against British and overseas standards, and space was made available, too, for examination of consumer goods on behalf of the then flourishing Consumer Advisory Council.

From the start it was recognized that the Hemel Hempstead Centre, since it was not involved in standards making, would not attract Government funds and must therefore be self-supporting. After a negligible loss in the early years, this aim was achieved, except in the case of the Technical Help to Exporters service which started as a wholly government-financed operation.

The story of the Centre has been one of steady growth and expansion to the present time, and one which reflects many of the major developments in BSI as a whole. At present, the Centre employs some 200 people, and has an overall annual expenditure of some £600 000 per annum. Its activities are divided between the test house – with its electrical, mechanical, physical and chemical, photometric and inspectorate sections – and the services provided to assist exporters. In many respects, however, the two parts of the Centre are complementary to one another.

Shortly after the setting up of the Centre, the need arose for a laboratory to test paraffin heaters to a revised standard. The urgency of the problem was such that the laboratory was constructed, installed and put into operation in some six weeks.

Mechanical engineering and electrical work continued to expand, and within 18 months of occupation of the Centre the first large building extension was put in hand to provide facilities for testing a wide variety of equipment, both for Kitemarking and in response to demands from individual manufacturers and organizations.

It was after the completion of this second phase of operations that agreement was reached with the National Physical Laboratory at Teddington to take over a number of projects involving routine testing that had formerly been operated by the NPL. The transfer of this work was phased over several years and included the testing of items such as clinical and industrial thermometers, volumetric glassware and hydrometers, and taximeters. A special feature of the work taken over from NPL is that, with few exceptions, the schemes involve testing the total production of items to be certified – all items are non-destructively tested to ensure compliance with the appropriate requirements. This work has increased in a number of cases: for example, the output of testing clinical thermometers has increased fivefold and now exceeds two million each year.

The publication of a standard for car seat belts created a need for a Kitemark scheme, and this was later embodied in legislation. The test equipment for this scheme was largely designed and produced in the Centre, at a time when few similar facilities existed anywhere else. One of the more spectacular pieces of equipment at the Centre is the dynamic rig built to simulate the effects of crash conditions on car seat belts, a rig in which a life-sized dummy figure on a steel sledge weighing half-a-ton and travelling at 65 km/h is crashed to a halt inside a pair of massive wedge-shaped steel plates.

The Centre has been increasingly used to develop individual tests for incorporation in British Standards, and, in the last five years, two long term development programmes have been in operation. The first of these was a three-year development programme for the production of performance tests

Car seat belts being tested at the BSI testing laboratories at
Hemel Hempstead. This well known Kitemark scheme has
made a major contribution to road safety (see page 86)

[*To face p.* 86

An aerial view of the Hemel Hempstead Centre showing the
Sherfield Building opened by Lord Sherfield in 1971, at the
bottom right of the complex (see page 85)

for mattresses and, at the end of this three year programme, a similar but continuous scheme has been arranged to meet the needs of the bedding industry.

The second of the development programmes was set up in response to requests from industry for the provision of performance tests for aluminium and steel windows. This programme, which has been equally successful in producing performance tests for incorporation in British Standards, will be followed by a further two-year programme.

A relatively recent approval scheme set up at the Centre is concerned with the performance of diesel engines operating road vehicles, with particular reference to the acceptable limits of smoke emission. This scheme has been accepted by practically all UK manufacturers and overseas manufacturers in countries such as the United States and Sweden.

Until fairly recently the testing of appliances for the British Electrical Approvals Board was carried out at their own test station, but the Centre has now accepted continuing commissions for carrying out the testing of certain categories of appliances such as sewing machines and soldering irons. At the present time a large volume of work is in hand for monochrome television sets which will be followed later by colour television, and ultimately by other types of equipment such as tape-recorders.

Work in all the testing sections at the Centre is steadily expanding and agency arrangements have been made with many organizations for carrying out on their behalf not only testing work but also factory surveillance. Typical of these agency arrangements are those with other national standards bodies, such as the Canadian Standards Association and the Standards Association of Australia, and with Government departments in the UK.

Technical help for exporters

One of the main aims of the Centre has always been the support of the export trade, and the service it renders individual manufacturers by assuring compliance of their equipment with the standards of various countries overseas has contributed in no small way to Britain's export successes.

Foremost among these contributions is the Technical Help to Exporters service, which was set up in 1966 with government support. Largely based in Hemel Hempstead, the service provides British exporters with speedy and authoritative technical guidance to help them through the tangled jungle of specifications and regulations relating to their products in overseas markets. THE staff give assistance in obtaining test certificates or approval for products destined for overseas markets, and arrange for testing and inspection in the the UK wherever possible. THE engineers have so far visited some 40 countries, and prepared over 200 digests of the technical requirements of individual countries, governing their imports of electrical equipment, cranes, boilers and pressure vessels, buildings and building components, and gas

87

equipment. Much of the credit for this imaginative scheme belongs to the Centre's director, Mr J P Roberts, the inspiration for it coming directly from his long experience in providing British exporters with guidance on the standard practices and rules of other countries. With Britain a member of the European Economic Community, the role of the Hemel Hempstead Centre in providing British industry with an export advisory service will prove even more significant.

Though the Centre started from a firm base of existing need, its establishment was also something of a ' leap in the dark ', as Lord Sherfield said when opening the building named in his honour in 1971 – the fifth extension to the Centre in 12 years. That leap has already been amply rewarded, and the future development of the Centre seems assured.

14

The role of BSI in a changing world

The beginning of the 1970s saw BSI in a strong position as a national body right at the centre of major economic and industrial advance. During the previous 20 years, under the direction of Mr Binney, the activities of BSI had expanded dramatically. When, in 1970, Mr G B R Feilden took over as the chief executive of BSI, it could be claimed that, directly or indirectly, British Standards affected the major part of industry's output.

But rapid growth poses problems more daunting than those of more steady progress. By 1971, partly as the consequence of its success in so many fields since the Second World War, BSI faced the cumulative problems of a period of exceptional growth, with little opportunity to pause and consolidate.

With its dependence on the support of industry for its lengthening programme of work, BSI was vulnerable to shifts in the economic climate, especially when the confidence of industry was shaken by events such as the collapse of Rolls Royce. Inflationary pressures, too, were pushing up costs at an alarming rate. BSI, operating labour intensive services in Central London, was particularly hard hit. World trading patterns were changing, and above all, with the prospect of entry to the European Common Market, loomed the uncertainty of what this would mean for British industry.

It was indeed timely that this was also the moment when progress in international standards activities culminated in several important agreements. In 1971 the decision to publish European Standards was taken by the 13 member nations of CEN; and in the same year it was agreed at the ISO General Assembly that ISO Recommendations would henceforth be termed International Standards. These two developments put the future of national standards work under the microscope. Increasing international work, involving participation in numerous overseas meetings, demanded the commitment of a mounting proportion of BSI's resources. On the other hand, national standards work was becoming more and more an endorsement of what had already been agreed internationally. To those who argued that BSI was spending too much on international standards projects, the simple answer was that it dare not fail to wield its maximum influence in representing the views and needs of British industry in the international forum.

The United Kingdom was not alone in this dilemma. In some countries which had hitherto concentrated their standardization efforts on national work, especially the United States and Canada, the problem was particularly acute, leading in 1971 to urgent government consideration in both countries about the means of increasing the level of their participation in international standards negotiations. In most countries, to a greater or lesser degree, there was an appreciation of a new relationship between national and international standards. Whereas in the late 1940s international agreement was a desirable follow-up to agreement reached nationally, by the 1970s international negotiations had become the major consideration affecting national standards work.

By 1971 BSI was particularly well-placed to reap the benefits of progress in the further harmonization of standards internationally. As the holder of a record number of ISO and IEC secretariats, and represented on nearly every international technical committee and working group as well as the major policy committees, BSI was at the centre of discussions in an ever widening range of subjects. At the beginning of 1972, BSI's Executive Board set up an International Policy Panel to advise on BSI's tasks in world standardization. Recognizing the different needs of different areas, BSI's attitude to regional developments in standardization and the trade groupings they serve was important: but what were to be the priorities? A greater need for communication and liaison arose, too, from the activities of United Nations agencies concerned with public health and safety.

With this surge of international work, and with the metrication programme at its peak, the gap between the work-load and the resources available began to widen, leading to new pressures and fresh problems. Some means had to be found of reconciling increasingly widespread and urgent demands on the Institution with limited staff resources. In particular, it was vital to ensure that priorities for different projects were properly assessed and reflected in the work programme.

In 1970, the latest in a series of self-appraisals conducted by BSI to examine how effectively BSI was fulfilling its role was contained in the report of the second committee under the chairmanship of Sir Anthony Bowlby set up to review BSI's procedures. The main recommendations of the committee were that more stringent control should be exercised over the acceptance of new projects, over the allocation of priorities, and over the setting of target dates and the methods of recording progress. The committee also stressed the desirability of preparing initial drafts by small panels, trade associations, or BSI staff, before submitting them to full technical committees. This particular enquiry made a careful study of criticisms contained in the report of the Ministry of Technology's committee on pressure vessels technology and standards in 1967.

The pressure vessels report raised other issues of central importance to BSI's future work. The suggestion that a new pressure vessels authority should be established was flatly rejected, but BSI accepted that there was a need to streamline committee structures and reconstitute them on the basis of technologies rather than industries. A new Industry Standards Committee

Members of the Executive Board of BSI, October 1969.
Left to right, standing: Brigadier R A Blakeway, Mr N F
Marsh, Mr K McLaren, Mr T W Howard, Mr G Weston
(adviser to BSI), Mr G B R Feilden, Mr K M Wood,
Mr P H Dunstone, Mr K C Hunt, Mr H A R Binney,
Mr A H A Wynn Mr S E Goodall. Sitting: Miss M F
Hardie (secretary to the Board), Miss E Ackroyd, Mr C H
Colton, Mr G H Beeby (chairman), Mr P L Cocke, Mr C
Needham, Mrs A Stanley, Lord Carron, Mr C G Cruick-
shank. Other members of the Board, not present at the
meeting, were: Sir Anthony Bowlby Bt., Mr C D Bunce, Mr
Geoffrey Cunliffe, Mr E W Greensmith, Mr J M Langham,
Mr A H Mortimer, Mr T A L Paton, Mr L W Robson

[To face p. 90

Mr G B R Feilden photographed during the recording of the
BBC television programme 'Metrication – the task for
management', broadcast in March 1972

was established for all pressure vessels work, and this was followed by similar moves in the instrumentation field with the establishment of an Industry Standards Committee for process control.

The problem of reconciling the requirements of rapid technological progress with those of long term agreement is a perennial one in standards work, and from time to time efforts have been made to solve it. As long ago as 1915, interim reports were issued for specifications such as the threads of sparking plugs which at that time were matters of considerable uncertainty. In 1970, a new type of publication, the ' Draft for Development ' was introduced as a formal approach towards reconciling standardization needs with technological advance. Where there was insufficient information or experience available, or lack of unanimity of view, a draft for development would provide a basis from which a British Standard could ultimately be developed. As well as providing a useful guideline, the draft would, it was hoped, stimulate constructive thinking and lead to further research.

The publication of drafts for development highlighted another key problem – that of securing adequate feedback from users to the relevant technical committees. Allied to this was the whole question of the design and utilization of British Standards. Throughout the history of the Institution, the presentation of British Standards had scarcely altered; yet techniques in handling, storing and retrieving technical information had undergone a revolution in this period. To facilitate the use of standards, a new A4 size format was agreed and, to take into account the problems of certain users, a new committee was established to study the computer implementation of British Standards. Arrangements were also made for increased participation by representatives of users, such as the Standards Associates, in the making of decisions of this nature.

A vital corollary to standardization, both national and international, is the existence of schemes for certification and approval. An important development in this field was the establishment in 1971 of a new Quality Assurance Council within the structure of BSI, following the publication of the Mensforth Report, which identified the need for a coordinating body. The task of the Council, representative of all major interests, is to promote the adoption of reliable certification schemes acceptable to manufacturers, Weights and Measures departments, and the purchaser.

Perhaps the greatest single challenge facing BSI at the start of the seventies was no less than a full reappraisal of its future relations as an independent grant-aided body with the three main constituent elements of its support: industry, government, and the consumer; the interests of each frequently needing reconciliation with those of the others (and the financial support from each not necessarily matching its interests). BSI's task has always been to find common ground for all interests, on which a solution acceptable to all parties can be based.

BSI has been particularly fortunate in the support for its activities that it has received from industry, directly and through the Confederation of British Industry and other representative bodies. Recent years have seen an increasing demand for standards concerned with the wider problems of the environment and the consumer, and while these portend an exciting future, this will only come about through the active support and participation of all sections of society in the work of the Institution.

The key to the success of BSI throughout its first 70 years has been its reputation, derived from the readiness of all to accept British Standards, and to provide their support for the operation of consensus procedures, both as standards users, and participants in technical committees. The achievement of a right balance of support that will enable the enormous workload in hand to be effectively processed and, at the same time, preserve the voluntary nature of standards work is likely to be one of the most pressing problems of the years ahead. It is, and always has been, difficult to quantify the immediate benefits that result from the implementation of standards. But the history of BSI shows clearly that an economy with the right standards makes the best possible use of its resources.

As a leading personality of the European Economic Community expressed it at a joint CBI/BSI Conference in 1971, 'Efficiency depends on having a set of standards which are modern, complete and non-discriminating.'

Such a set of standards reflects the economic and industrial development of a country: if they are *modern*, it means that industry is producing technologically advanced products; if they are *complete*, it means that the country is functioning in all sectors of the economy; if they are *non-discriminating*, it indicates that the country has decided to face the challenge of free competition.

Appendix 1

The growth of BSI

Year	New and revised standards issued	Numbers of committee meetings	Copies of standards distributed	Income from sales £
1910	3	n.a.	n.a.	1688
1920	77	350	63 000	1633
1930	52	850	n.a.	1970
1940	93	490	157 000	11 299
1950	194	3060	656 000	52 837
1960	301	5030	1 158 000	206 709
1970	630	9530	1 818 388	715 890

Year	No. of sub-scribing members	Income from subscribers £	Government grant £	Total budget £
1910	120	1055	800	3731
1920	1500	10 350	4166	16 200
1930	n.a.	12 876	4207	26 852
1940	2359	17 167	7070	38 000
1950	6306	78 721	90 000	234 065
1960	10 697	209 456	170 000	701 246
1970	14 561	561 452	1 137 025	2 829 645

n.a.=information not available

Appendix 2
Some key figures in the history of BSI

Sir John Wolfe Barry KCB FRS (1836–1918) was Chairman of the Engineering Standards Committee from 1905 until his death. A distinguished consulting engineer, his projects included Tower Bridge and other Thames bridges, the London underground Circle railway, and the Barry docks in South Wales. He also served on a number of royal commissions. His devotion to the work of the Committee ensured its official recognition and incorporation as the British Engineering Standards Association in 1918.

Charles Le Maistre CBE (1874–1953) succeeded Leslie Robertson as Secretary of the Engineering Standards Committee in 1916. He held the chief executive post until 1942, thus becoming the first Director of BSI. As General Secretary of IEC from its inception in 1904 until his death, he played a vital role in the development of international co-operation in the expanding field of electrical technology, and pioneered the concepts of international harmonization.

Percy Good CBE (1880–1950) was appointed Assistant Secretary of the Committee in 1914, and was appointed Deputy Director of the Association in 1929 and Joint Director of BSI in 1941. From 1942 until his death he was sole Director. He was President of the Illuminating Engineering Society in 1938/9, and of the Institution of Electrical Engineers for the year 1947/8. The success of BSI work during the 1939–1945 war, and its subsequent expansion, were largely due to his vision and energy.

Sir Roger Duncalfe (1884–1961) was President of BSI 1953–6 and of ISO 1956–8. Having risen rapidly to become Managing Director of his family firm, he organized and became Chairman of British Glues and Chemicals Ltd in 1920. Prominent in numerous trade associations and committees, he was a Vice-President of the FBI and Deputy Chairman of the Beaver Committee on Air Pollution.

Mr G Weston CBE began his engineering career with British Thomson-Houston and later went to the Fuel Research Station where he was engaged in early work on atmospheric pollution, particularly in relation to smoke emission from industrial boilers. He joined the British Engineering Standards Association in 1927 and subsequently became Assistant Director of BSI in 1948, Technical Director in 1950, and Associate Director from 1960

Mr Percy Good

Mr G Weston

To face p. 94

Mr H A R Binney

Mr G B R Feilden

until his retirement in 1969 after 42 years' service. From 1969 to 1971 he acted as Adviser to BSI. Mr Weston has represented the UK overseas on many occasions, chaired numerous international meetings, and assisted in the setting up of standards organizations in the Commonwealth. During the formative years of ISO he helped plan the staff structure of the central office in Geneva and advised on new procedures.

Mr H A R Binney CB, the chief executive of BSI, first called Director and then Director General, for the twenty years 1951 to 1970. Now the Adviser, International. He spent the first half of his career with the Board of Trade. He was Controller of the Import Licensing Department at the outbreak of the 1939–1945 war and was head of the first Engineering Division of the Board of Trade formed in 1944 with the main object of facilitating the transfer of industry from war to peace production. In 1947 he became Under Secretary responsible for Government interests in relation to all textile and clothing industry matters. As chief executive of BSI, Mr Binney guided the organization through a key period in its development. He has been Vice-President of ISO and Chairman of CEN and a leading member of ISO and IEC policy committees. He has travelled extensively on ISO and IEC affairs and is currently Chairman of the recently established ISO Committee called CERTICO concerned with quality assurance matters.

Mr G B R Feilden CBE FRS, the present Director General of BSI, started his distinguished career in engineering with Lever Brothers Ltd, before being transferred to Sir Frank Whittle's team to work on jet propulsion. After the 1939–1945 war, he joined Ruston and Hornsby, becoming Engineering Director in 1954. In 1959, he was appointed Managing Director of Hawker Siddeley Brush Turbines Ltd., and in 1961 became Group Technical Director of Davy-Ashmore Ltd. Mr Feilden was a Vice-President of the Royal Society, and has served on several scientific councils, including the DSIR Committee on Engineering Design which reported in 1963, of which he was Chairman. He was appointed Deputy Director General of BSI in 1968, and succeeded Mr Binney as Director General in 1970.

Appendix 3

Presidents and Chairmen

President	The Lord Sherfield, GCB, GCMG
Chairman of Executive Board . .	Mr E W Greensmith, OBE
Deputy Presidents	{ Mr G H Beeby { Sir Anthony Bowlby, Bt
Chairman of Finance Committee . .	Mr T W Howard

PAST PRESIDENTS

The Rt Hon. Lord Woolton, CH	1944–1947
The Rt Hon. Lord McGowan, KBE, DCL, LLD . . .	1947–1949
Sir William Larke, KBE	1949–1950
The Rt Hon. Viscount Waverley, GCB, GCSI, GCIE, FRS .	1950–1953
Sir Roger Duncalfe	1953–1956
Sir Herbert Manzoni, CBE	1956–1958
Mr R E Huffam, MC	1958–1961
The Hon. Geoffrey Cunliffe	1961–1963
The Rt Hon. the Earl of Kilmuir, PC, GCVO . . .	1963–1966
Sir Anthony Bowlby, Bt	1966–1967

PAST CHAIRMEN

Mr James Mansergh	1901–1905
Sir John Wolfe Barry, KCB, FRS	1905–1918
Sir Archibald Denny	1918–1927
Mr Maurice F G Wilson	1928–1933
Mr E J Elford	1933–1934
Dr E F Armstrong, FRS	1934–1935
Mr W Reavell (*later Sir William Reavell*) . . .	1935–1936
Mr E J Elford	1936–1937
Dr E F Armstrong, FRS	1937–1938
Sir Frank Heath, KCB, CBE	1938–1939
Sir Percy Ashley, KBE, CB	1939–1944
Sir William Larke, KBE	1944–1947
Sir Clifford Paterson, FRS	1947–1948
Sir Roger Duncalfe	1948–1952
Mr John Ryan, CBE, MC	1952–1955
Sir Herbert Manzoni, CBE	1955–1958
Mr R E Huffam, MC	1958–1961
The Hon. Geoffrey Cunliffe	1961–1964
Sir Anthony Bowlby, Bt	1964–1967
Mr G H Beeby	1968–1970

Note. From 1956 to 1963 and in 1966–67 the offices of President and Chairman were combined. The Presidency was not filled in 1968 and 1969.

Index

Local Authority Standards Advisory Committee 81

M

Macmillan, Harold 19
Malaysia 50
Malta 51
Management, British Institute of 14
Martin, Bruce 20
Martin, C A J 42
Martin, Professor Harold W 37, 62
Mauritius 51
Medical Association, British 29
Medical equipment, standards for
 clinical thermometers, testing of 86
 hospital equipment 5,16
 hypodermic syringes and needles 56
Mensforth Report 30, 84, 91
Mensforth, Sir Eric 30
Methods of test 3-4, 29
Metrication 5, 30, 31, 32-33, 42-47, 51
 construction programme for 44
 engineering programme for 45
 Government decision on 43
 key symbol 46
 SI units 3, 45, 53
 White Paper on 47
Metrication Board 46
Metrication, Standing Joint Committee on 46
Modular design (*see also* dimensional co-ordination) 27
Modular Society 20
Molony Committee 74, 80
 Report 29, 74
Molony, J T 26, 74

N

National Council for Quality and Reliability 83, 84
National Council of Social Science 75
National Council of Women 69
 Advisory Committee on Consumer Goods 69
National Physical Laboratory 86
Natural gas 29
Needham, Christopher 30
Netherlands 39, 65
New Zealand, Standards Association of (SANZ) 46, 50
Nigeria 51
Nixon, F 83
Noise 5, 56
 measurement of vehicle 28
Norway 53, 63
Nuclear energy 5, 22, 24, 56, 58

O

Oil Appliance Manufacturers Association 79, 81

Oil Burning Equipment Testing Association, Domestic 82
Oils, lubrication 40
Organization for European Economic Co-operation (OEEC) 16, 62
 Committee on Productivity and Applied Research 62
Ottawa
 Standards conference (1945) 52
 Standards meeting (1962) 51

P

Packaging 13, 22, 24
 code 14
 retail trade quantities 65
 standards for:
  freight containers 57
  medicines 14
  pallets 15, 28
Paper sizes 23, 91
Pearson, Professor E S 83
Petroleum 21, 64
Pipes
 cast iron 9
 copper tubes, testing of 79
 PVC 79
Plastics 15, 79
Plastics Federation, British 78, 81
Pollution 5, 22
 Beaver Committee on 94
 legislation for 22
 standards for:
  smoke density indicators 22
  smoke emission tests for diesel engines 87
  smoke viewers 22
Portugal 63
Pressure vessels 21, 31, 87, 90-91
 committee on 90
 regulations for 65
Process control 59, 91
Production, Ministry of 14
Productivity Council, British 37, 38
Purchasing, public 9, 16, 17, 44, 65, 81
 White Paper on 81

Q

Quality assurance (*see also* certification and Kitemark) 4, 5, 30, 70, 84
 quality control in 12-13, 14, 80, 82-84
Quality Assurance Council 30, 84, 91

R

Rails, tramway 9, 78
Railways
 Indian 49
 locomotives 8, 9
 rolling stock 9, 10